BEHAVIOURAL AND
ECOLOGICAL GENETICS

BEHAVIOURAL AND ECOLOGICAL GENETICS
A study in *Drosophila*

BY

P. A. PARSONS

Professor of Genetics, La Trobe University, Melbourne
and Sometime Fellow of St. John's College, Cambridge

CLARENDON PRESS · OXFORD
1973

Oxford University Press, Ely House, London W.1
GLASGOW NEW YORK TORONTO MELBOURNE WELLINGTON
CAPE TOWN IBADAN NAIROBI DAR ES SALAAM LUSAKA ADDIS ABABA
DELHI BOMBAY CALCUTTA MADRAS KARACHI LAHORE DACCA
KUALA LUMPUR SINGAPORE HONG KONG TOKYO

© OXFORD UNIVERSITY PRESS 1973

PRINTED IN GREAT BRITAIN
BY BUTLER & TANNER LTD, FROME AND LONDON

PREFACE

In general, genetics, ecology, and the study of behaviour have been regarded as separate entities. This is not realistic in the development of a coherent evolutionary biology of an organism, and we should look towards a unification of all three. While with time this will inevitably occur, the existing trichotomy occurs because geneticists, ecologists, and behaviourists usually have somewhat different approaches. The geneticist tends to choose clear-cut genetic markers or quantitative traits which can be analysed relatively easily in order to establish patterns of variation. He may search as a by-product for behavioural or eological correlates. The behaviourist tends to choose traits representing different types of behaviour without necessarily looking for ecological or genetical consequences, and the ecologist tends to look at traits of obvious adaptive significance, some of which may have behavioural components, but he may only consider their genetic basis at a secondary level.

For many organisms we lack the necessary background information for the unification of all three entities. In the genus *Drosophila* there have been a number of studies linking two of the entities, but the linking of all three has been little attempted. However, when considering the distribution of species of this genus, and genotypes and races within species, it is clear that the influence of behavioural and ecological factors must be considered in order to understand the genetic architecture of the species. This book is an attempt at such an integration using the genus *Drosophila*. The principles discussed should be applicable to many other genera, especially those relatively close to *Drosophila* phylogenetically. After a brief introduction to the genus, its behaviour genetics is discussed, followed by its ecological genetics, since some of the topics in ecological genetics have behavioural components. In the last part of the book, an attempt is made at integrating the earlier parts in the hope of contributing to the general evolutionary biology of the genus, and possibly to evolutionary biology as a whole. The literature review upon which the book is based ceased in mid-1971.

It is my impression that there will be substantial developments in the integration of genetics, behaviour, and ecology in the next few years. It is hoped that this book will be of some help in indicating some of the parameters that need to be taken into account. Needless to say, the mode of attack would be very different in organisms distant from *Drosophila* phylogenetically, but even so, there may be some points of relevance. It was felt best to restrict the discussion to the one genus, since by so doing it was possible to aim at a coverage in some depth; also it is a genus with which I have some familiarity.

In any book of this nature, assumptions must be made about the levels of knowledge of readers. An elementary knowledge of genetics and statistics is assumed; certain principles taken from the fields of population and biometrical genetics are explained briefly. No attempt is made at providing an account of the general biology, genetics, and cytology of the genus, as this has been the subject of numerous reviews and books. However, in the first chapter some indication is given of the extensive contributions made by those working with the genus to biology as a whole. The contributions are substantial, and it is my belief that this trend will continue as new biological fields develop in the future.

I wish to thank Professor D. G. Catcheside, F.R.S., of the Research School of Biological Sciences, Australian National University, who helped me to convince myself that this project should be attempted.

I am grateful to Mr. J. A. McKenzie who read and criticized the whole manuscript, and to Dr. I. T. MacBean and Dr. N. D. Murray who read and criticized the behavioural and ecological sections respectively. I wish to acknowledge the secretarial help of Miss Glenda Wilson, and the assistance given by Mrs. C. Greer with the diagrams. Thanks are due to the authors, editors, and publishers of the following works and journals for permission to use published figures (for which appropriate reference is given in each caption): F. J. Ayala, *Canadian Journal of Genetics and Cytology*; M. Bastock, *Evolution, Lancaster, Pa.*; J. F. Crow, *Annual Review of Entomology*; Th. Dobzhansky, *Heredity, London*; the editor, *Genetics, Princeton*; K.-I. Kojima, J. Gillespie, and Y. N. Tobari, *Biochemical Genetics*; R. C. Lewontin and Y. Matsuo, *Proceedings of the United States National Academy of Sciences*; T. Narise, *Evolution, Lancaster, Pa.*; E. B. Spiess and B. Langer, *Proceedings of the United States National Academy of Sciences*; J. A. Thomson, *Canadian Journal of Genetics and Cytology*; B. Wallace, *Topics in population genetics*, Norton, New York, and the *American Naturalist*, University of Chicago Press.

Melbourne, 1971,   P.A.P.

CONTENTS

1. Introduction: Why choose *Drosophila*?	1
PART I: BEHAVIOUR GENETICS	
2. Behaviour genetic analysis: introductory	7
3. Genes with behavioural effects	11
4. Inversion sequences and mating behaviour	31
Appendix: Equilibria for a locus with two alleles A and a having different fitnesses	39
5. Quantitative traits	41
6. Selection experiments	61
7. Deviations from random mating	78
PART II: ECOLOGICAL GENETICS	
8. Genetic heterogeneity for environmental stresses	90
9. Competition	110
10. Dispersion and migration	118
11. Population size	127
12. Distribution of genotypes within species	136
13. Environments, enzyme variants, and genetic architectures	152
PART III: SYNTHESIS	
14. Distribution data between species	162
15. Behavioural and ecological isolation	178
16. Behaviour, ecology, and evolution	186
Bibliography	193
Author Index	217
Subject Index	221

1
INTRODUCTION: WHY CHOOSE DROSOPHILA?

A major reason for choosing *Drosophila* is that much of the chromosome theory of heredity was based on *D. melanogaster*, commonly called the fruit-fly or vinegar-fly. Its great advantage is its ease of cultivation in the laboratory. Within two to three weeks, a single pair can provide several hundred offspring which can be observed at any stage; eggs, three larval instars, pupae, and adults. Differences in morphology, pigments, viability, behaviour, and ecological tolerances can easily be assessed, and in recent years a multiplicity of enzyme and protein variants have been found. A great amount of general information about the biology of the genus may be found in Demerec (1955).

T. H. Morgan in late 1909 began to use *D. melanogaster* in his laboratory; beginning with an attempt to detect induced mutations, he discovered sex-linkage in 1910, using white-eyed flies. Before long further sex-linked mutants were found, and by 1911 he put forward a theory of recombination. Then followed a period of intense activity by Morgan and his associates, Sturtevant, Bridges, and Muller, which rapidly led to a chromosome map of *D. melanogaster* which is probably now far more complete than that for any other known higher organism. The extraordinarily rapid expansion of knowledge of the physical basis of heredity in the decade 1910–19 was due in large measure to the thorough and detailed study of this one species. Bridges, in particular, spent a great deal of time developing and maintaining stocks for special purposes, and at the time of his death in 1938 some 900 such stocks were being maintained. The use to which some special stocks have been put in enabling the genes controlling behavioural and physiological phenotypes to be located to specific chromosome regions will become apparent in later chapters.

Drosophila has proved to be remarkably versatile in the development of modern genetics. As research trends changed, the insect was able to provide evidence even if the trend originally concerned organisms widely divergent from it. Beginning with its role in the elucidation of the mechanism of Mendelian heredity, there followed a detailed correlation of genetical and cytological behaviour, analyses of mutation, and, more recently, fine structure analyses. Major steps in physiological and developmental genetics have come through *Drosophila*, as have examples of non-chromosomal heredity on the boundary line between infection and heredity.

Much of the laboratory work aimed at verifying the theories of population genetics has been carried out with *Drosophila*, beginning with Chetverikov (1926) who found a great amount of hereditary variation in natural populations of *D. melanogaster*. This was but a forerunner of a vast amount of classic experimental work by Dobzhansky and his co-workers in *D. pseudoobscura* and other species which has continued from the 1930's to this day. Much of this work was based on polymorphisms for chromosomal inversions which segregate in the wild, and from this much has emerged on the genetics of populations in the wild. As well as this, some of the simpler theoretical models of population genetics were tested and found to be relatively realistic. The emphasis, however, is now changing since in the last decade, beginning with the development of starch-gel electrophoresis, a vast range of enzyme and protein variants segregating in natural populations have been found. Population genetics is now beginning to find its niche in the wider field of evolutionary biology, where an attempt is being made to understand all the processes determining the distribution and abundance of species in the wild. With this change of emphasis, the classical models of population genetics, elegant as they are, are now proving to be too simple. One cause of complication is behavioural, especially factors associated with mating, since this has led to the discovery of mating patterns difficult to analyse mathematically. Further complications occur at the ecological level; for example, fitnesses of karyotypes in *D. pseudoobscura* are often temperature-dependent. The classical models do not take these complications into account: at the present stage many models exist at the theoretical level, but biological reality is turning out to be more complex. Fairly sophisticated experiments are needed to characterize these complexities in the laboratory. These and other similar issues are discussed later: it will also be seen that behavioural factors as well as ecological factors are of importance in maintaining the discreteness of populations within species, and also of the species themselves. In fact, where studied, isolating mechanisms between sibling species have turned out to be a complex of behavioural and ecological factors.

Although information on the behavioural and ecological genetics of *Drosophila* has been appearing for many years, mainly as a by-product of other investigations, the emphasis in the past few years has changed more to a direct investigation of the behavioural and ecological genetics of *Drosophila*. Even so, behavioural and ecological genetics have been developing on the basis of studies in other organisms, perhaps at stages in advance of studies in *Drosophila*. For example, work on rodents has contributed over the years to behaviour genetics (see Fuller and Thompson 1960). Much of the earlier work in ecological genetics was on plants, butterflies, and land snails (see Dobzhansky 1951, Ford 1964, Clausen, Keck, and Hiesey 1940, 1948). Unfortunately, in many cases knowledge of the formal genetics of many of these species is restricted, as compared with *Drosophila*, but they do possess the advantage of readily identifiable phenotypes in the wild as assessed by visual differences. This is one advantage that *Drosophila* lacks, since, in the field, species of the genus are remarkably uniform in colour and form. In recent years, however, this disadvantage is being overcome because of the large number of enzyme and protein variants now being found in natural populations and for which single flies can be assayed. Furthermore, the ease of breeding of certain of the more cosmopolitan species has meant that an enormous amount of work has been published on *Drosophila* over the years, and the total amount of work that has appeared contributing to behavioural and ecological genetics is by now quite substantial. Therefore, this genus

TABLE 1.1

The present taxonomic arrangement of the genera and subgenera of the Hawaiian Drosophilidae

SUBFAMILY AMIOTINAE

Gitonides Knab,	1 introduced species
Pseudiastata Coquillett,	3 introduced species (not established)

SUBFAMILY DROSOPHILIDAE

Drosophila-like genera (drosophiloids)

Genus *Antopocerus* Hardy,	9 endemic species
Ateledrosophila Hardy,	2 endemic species
Celidosoma Hardy,	1 endemic species
Chymomyza Czerny,	1 introduced species
Dettopsomyia Lamb,	2 introduced species
Drosophila Fallén	
Subgenus *Drosophila* Fallén,	305 endemic species, 10 introduced species
Sophophora Fallén,	3 introduced species
Engyscaptomyza Kaneshiro	6 endemic species
	324 total
Grimshawomyia Hardy,	2 endemic species
Nudidrosophila Hardy,	5 endemic species

Scaptomyza-like genera (Scaptomyzoids)

Genus *Scaptomyza* Hardy

Subgenus *Alloscaptomyza* Hackman,	8 endemic species
Bunostoma Malloch,	8 endemic species
Exalloscaptomyza Hardy	6 endemic species
Parascaptomyza Duda,	1 introduced species
Rosenwaldia Malloch	6 endemic species
Tantalia Malloch,	6 endemic species
Trogloscaptomyza Frey,	86 endemic species
	121 total
Genus *Titanochaeta* Knab,	11 endemic species
Total	475 presently known species in Hawaii

After Carson, Hardy, Spieth, and Stone 1970

has been selected as a 'type' genus which should illustrate many of the problems existing in other organisms.

Much research in *Drosophila* has been restricted to a few species which can be easily handled in the laboratory. However, the world fauna contains between one and two thousand species. Of these, it is remarkable that up to 500 identified species occur in this and closely related genera (Drosophiloids) in the Hawaiian Islands, and there are probably 200 or more species belonging to the genus *Scaptomyza* and related genera (Scaptomyzoids) which also belong to the family Drosophilidae (Carson, Hardy, Spieth, and Stone 1970) Table 1.1 summarizes the present taxonomic arrangement of the genera and subgenera of the Hawaiian Drosophilidae (Carson *et al.* 1970). Many intriguing problems are emerging. One feature is the definite intergradation between the two major genera *Drosophila* and *Scaptomyza* in Hawaii, since elsewhere in the world they are differentiated. The number of species of *Scaptomyza* in Hawaii is twice that for the rest of the world. Detailed studies of the Hawaiian *Drosophilidae*, which are now beginning to reap rewards, should be of considerable significance in the evolutionary biology of the whole family, and in particular of the genus *Drosophila*. This also involves the study of the karyotypes of the various species and their evolution. Table 1.2, after Clayton (1968), gives a comparison of Hawaiian and non-

TABLE 1.2

Comparison of chromosome numbers and shapes from Hawaiian and non-Hawaiian members of the subgenus Drosophila

Haploid chromosome number	Non-Hawaiian species (no.)	(per cent)	Hawaiian species (no.)	(per cent)
Seven	1	0·7	0	0
Six	81	54·0	89	95·6
Five	36	24·0	2	2·2
Four	29	19·3	2	2·2
Three	3	2·0	0	0
Number of species	150		93	
Chromosome shapes†				
5R, 1D	53	35·3	80	86·0
6R	7	4·7	7	7·4
3R, 1V, 1D	11	7·3	2	2·2
1R, 2V, 1D	15	10·0	2	2·2
Other	64	42·7	2	2·2
Number of species	150		93	

† D = dot, R = rod, V = V-shaped chromosome.

After Clayton 1968

Hawaiian species of the subgenus *Drosophila*. This shows a preponderance of five-rods one-dot karyotypes in the Hawaiian species (86 per cent), and a fair proportion of these in the non-Hawaiian species (35·3 per cent). The preponderance of the five-rods one-dot karyotypes in Hawaii is significant in view of Patterson and Stone's (1952) conclusion that this is a primitive karyotype. These authors conclude that alterations of this primitive karyotype are the result of either fusion of whole arms with loss of centromere, pericentric inversion, or added heterochromatin. The genus *Scaptomyza*, on the other hand, has twelve of eighteen species tested with $n = 5$ (one V-shaped chromosome, three-rods and one-dot). Thus the subgenus *Drosophila* and the genus *Scaptomyza* tend to fall into two groups chromosomally in Hawaii, although with some intergradation. The general problem of karyotypes and their evolution is not discussed in this book because of its adequate coverage elsewhere (see, for example, Patterson and Stone 1952, Carson *et al*. 1970).

The Hawaiian species presumably represent an adaptive radiation of the Drosophilidae in an analogous way to Darwin's finches on the Galapagos Islands (see Dobzhansky 1968), probably arising from the chance arrival of one or two species. It is to the Hawaiian species that we must look for increasing results in the future, and an excellent start has been made by Carson *et al*. (1970). Of the Hawaiian species, Stone, Guest, and Wilson (1960) estimate that about 95 per cent are rare, and are geographically very localized, as well as being behaviourally and ecologically very specialized. In some cases their breeding and feeding sites are relatively well known, so that interrelationships between their genetic and ecological structure can be established (Carson *et al*. 1970). This tends to be more difficult for the more widespread and cosmopolitan species which occupy more diverse habitats. On the other hand, there are acute problems in breeding many of the Hawaiian species because it is difficult or impossible to simulate their nutritional needs in the laboratory. Therefore, although most of the work so far published has been on species easy to breed in the laboratory, these are just the species for which we have little information in the wild, and the converse seems largely true for many of the Hawaiian species. With time, it should be possible to converge, as with research more exotic species will no doubt be bred in the laboratory, so facilitating studies on their genetic architecture, and at the same time those species which are easy to breed in the laboratory will be studied progressively more from the point of view of their behaviour and ecology in the wild.

There are two main approaches to the behavioural and ecological genetics of *Drosophila* which will be followed: first, the study of flies in natural environments in order to correlate species, and genotypes within species, and their behaviour, with the environment; and second, the study of populations in artificial laboratory environments in order to assess what may be factors of behavioural and ecological importance in nature. As will be seen, work on these two aspects has proceeded rather independently, and the second approach has been considered in greater depth than the first because of the ease of breeding large numbers of *Drosophila* in the laboratory. It is, however, essential to link both approaches, and a start has been made on this. Throughout, examples of both approaches are discussed, but unfortunately much of the work consists of referral of the results of laboratory experiments subsequently back to their likely significance in wild populations. Both approaches are necessary for the understanding of the evolutionary processes leading to the genetic architecture of the various species of *Drosophila*, since it is essential to define the complex of interacting systems making up the ecosystem, and hence to define the niche the species would occupy within that ecosystem. Therefore, behavioural and ecological factors either known to be, or likely to be, of importance are discussed in relation to *Drosophila*, and an attempt is made to integrate these in the building up of an evolutionary biology of the genus, bearing in mind that the principles in a modified form should be applicable to many other genera, although the level of modification will necessarily depend on the phylogenetic position of the other genera relative to *Drosophila*.

PART I

2

BEHAVIOUR GENETIC ANALYSIS: INTRODUCTORY

THE genetic analysis of behaviour of some species of *Drosophila*, in particular *D. melanogaster*, has proceeded to a fair level of sophistication. There are two main prerequisites for this.

(1) The genome selected must be well known and well studied with plenty of useful marker stocks. *D. melanogaster* in particular, and to a lesser extent certain other species (for example, *D. pseudoobscura*), fulfil this criterion perhaps better than any other species so far studied from the behaviour genetic point of view. Associated with this is a need to be able to breed the species easily and to develop special strains, such as inbred strains, which is relatively easy in *D. melanogaster*. Unfortunately, these criteria cannot be fulfilled by some of the rarer and more specialized species because little is known about their genomes, and breeding in the laboratory, if possible at all, tends to be difficult.

(2) It is essential that the behavioural trait can be assessed objectively, which, for the common species of *Drosophila*, does not present great difficulties in the laboratory. In higher organisms, where learning and reasoning become progressively more important culminating in man, objectivity becomes progressively more difficult to attain. Even in mice, many forms of behaviour depend on previous experience; for example, the success in mating of some strains of mice depends on whether the sexes are reared together or in isolation (see Parsons 1967a). While isolated reports indicate some effects of previous experience in *Drosophila*, it is undoubtedly less important than in rodents and man.

For behavioural traits, in many cases the phenotype is fairly distantly linked with the genotype. A few behavioural traits are more directly under the control of single genes, but they are generally more remote from the primary gene products than are biochemical or physiological traits, and to a lesser extent morphological traits. There is sometimes a direct dependence of behaviour on variations in these traits, but often the connexion is tenuous, although detailed work can alter the situation. Because of these rather indirect links between behavioural phenotype and genotype, many behavioural traits are controlled by several genes and their interactions, so making the phenotype quantitative rather than qualitative. Furthermore, at all the various steps leading to the phenotype, environmental variations may play a part. This again represents a tendency of the behavioural trait towards becoming quantitative rather than qualitative. If meaningful inferences are to be made about the genetic control of quantitative traits, completely objective measurement is essential, as stressed earlier. Strict environmental control is also necessary, and as will be seen later, there is plenty of evidence for variations in behaviour due to the environment. These points all apply to morphological and physiological traits, but behavioural traits provide greater difficulties. Reasons of this type probably explain why the study of the genetic control of behaviour has lagged as compared with that of the genetic control of biochemical, physiological, and morphological traits, since rather sophisticated experiments may be needed to separate clearly the influences of genotype and environment, and to assess the importance of their interactions. These complications probably form a major reason why definitive work on the genetic control of behaviour has been restricted to a few organisms, and of these *D. melanogaster* is of considerable significance. Caspari (1968) regards behavioural traits as similar to those of plants so far as the effect of environment is concerned. In comparison, morphological traits in animals are much less sensitive to the environment. Therefore, it is important to look to some of the techniques of plant biometrical genetics for the analysis of behavioural traits.

The discussion of the genetic analysis of behavioural traits in *Drosophila* is based on three main approaches. The first is by the analysis of the behavioural effects of single genotypes, usually mutant genes. These are the simplest and easiest types of behavioural variations to investigate, but such genes are often rare and deleterious. Hence, from the population point of view they may not be of great importance, but because the variation they control is discontinuous, the genotypes they control may provide information on the types of behavioural variations that occur in populations, and which are open to investigation by biometrical methods. As well as behaviour controlled by rare genes, there are behavioural variants associated with genotypes making up a polymorphism. A good example concerns aspects of sexual behaviour in species such as *D. pseudoobscura* which depend on the karyotype. Such variations could be of considerable significance in the population.

Unfortunately, much variation is not under the control of specific loci or inversions that can be directly assigned to specific positions on chromosomes, but is quantitative rather than qualitative. This means that it is impossible to split the population under study into discrete groups for the trait, and so the variation observed is continuous. In order to analyse such variation, the methods and techniques of biometrical genetics must be used, the basic aim being to assess what proportion of the total variance of a trait, the phenotypic variance, can be assigned to genotype and environment. A number of techniques are available, many of which have been used in plant-breeding work; some of these have also been used in behaviour genetics research in recent years. This second approach, however, being essentially statistical throws little light on the basis of variability at the chromosomal and genic level. Such methods are based on a number of assumptions; thus, it is commonly assumed that genes controlling a quantitative trait produce an effect which is small and additive. Unfortunately, the simplifying assumptions of biometrical genetics cannot be substantiated from the few studies where the actual genetic architecture of quantitative traits has been worked out (see Lee and Parsons 1968). Neither can the simplifying assumptions take much account of the various diverse types of interactions possible between loci. Perhaps the most difficult problem of all is the question of a scale on which a trait is measured, since interactions can be radically altered by transforming the data. Moreover, interactions between genotype and environment pose severe problems which are almost intractable.

These limitations do not mean that the biometrical approach has no value. In fact, to obtain a preliminary idea of the relative importance of genotype and environment, some biometrical procedures are often carried out. Particularly useful for behavioural traits is the diallel cross, which is the set of all possible matings between several strains or genotypes. If inbred strains are used, inferences about the genetic control of the trait can be made in some cases.

The third approach to genetic analysis comes from the directional selection experiment, in which individuals are selected at the high or low extremes of a distribution in the hope of forming separate high or low lines in subsequent generations. Provided that a trait has some genetic basis there should be a response, since by selecting extreme phenotypes, extreme genotypes will be selected. A theory of the short-term effects of selection has been available for some time, and is mainly based on the principles of biometrical genetics. However, this cannot predict the ultimate response to selection (Lee and Parsons 1968). The advantage of the selection experiment from the point of view of genetic analysis is that the extreme phenotypes continuously favoured during selection will be controlled by genotypes which are likely to be more homozygous than the unselected base population. In the species *D. melanogaster*, in particular, by the use of some of the special stocks available, it is possible to analyse the response to selection to the chromosomal, and in some cases the genic, level. In theory it is possible to obtain detailed information on the genetic architecture of a trait from the selection experiment, as has been done by Thoday (1961) and his co-workers for sternopleural chaeta number (see Lee and Parsons 1968 for references to this and other work). Such work has revealed the invalidity of some of the assumptions of biometrical genetics. Compared with biometrical techniques such as the diallel cross, the selection experiment approach has the disadvantage of requiring many generations of work. However, because of its short generation interval, *D. melanogaster* lends itself to this approach for quantitative traits, including behavioural traits. In higher organisms such as mice the time needed would be prohibitive. Therefore, from the point of view of studying the actual genetic architectures of behavioural traits, it is unlikely that *Drosophila* will be displaced for many years. In Chapters 3–7 these various approaches are discussed, and some of the published work is reviewed; evolutionary aspects are considered especially in Chapter 7.

3
GENES WITH BEHAVIOURAL EFFECTS

1. Yellow and wild-type

IN 1915 Sturtevant noted that yellow males were usually unsuccessful compared with wild males when competing for females. It seemed possible that this was because the behaviour of yellow males differed from that of wild males, making them less stimulating to the females. Bastock (1956) studied the detailed mating behaviour of yellow and wild-type males. The wild stock was crossed to the yellow stock for seven generations, so that the wild stock became genetically similar to the yellow stock except in the region of the yellow locus. All the observations were made on pair matings for flies aged 4–5 days, the males and females having been kept in separate vials after emergence for this time. The percentage successes in one hour are given in Table 3.1 The yellow × yellow percentage success is far lower than

TABLE 3.1

Percentage success in one hour from pair matings using yellow and wild-type D. melanogaster

	Before crossing the wild stock to the yellow stock for 7 generations	After crossing the wild stock to the yellow stock for 7 generations
Wild ♂ × wild ♀	62	75
Yellow ♂ × wild ♀	34	47
Wild ♂ × yellow ♀	87	81
Yellow ♂ × yellow ♀	78	59

After Bastock 1956

for wild × wild. In considering matings between yellow and wild-type, the yellow male × wild female percentage success is lower than the wild male × yellow female. Thus, in the crosses where there are yellow males the percentage success is much lower than where there are wild-type males. The success of a given female therefore depends on the type of male. Hence the difference in behaviour seems to reside largely in the reduced mating success of the yellow male as compared with the wild-type male.

Bastock analysed courtship behaviour into three components, namely orientation, vibration, and licking. *Orientation* occurs at the beginning of courtship and involves the approach and the following of the female by the male. *Vibration* is the wing display of the male after orientation; it is repeated at short intervals followed by periods of rest. The wing vibrated is usually the one nearer the head of the female, who receives the stimulus via her antenna. The importance of the male wing display and its perception by the antenna of the female can be shown by studying the mating success of combinations of winged and wingless males, and females with and without antennae. Bastock (1956) showed that wingless males are much less successful than normal males when mating with normal females (Fig. 3.1). When the females were deprived

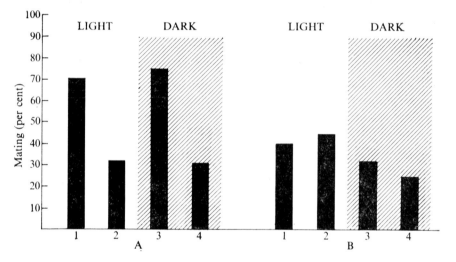

FIG. 3.1. Percentage of matings in one hour for winged (1, 3) and wingless (2, 4) males, and normal (A) and antenna-less (B), females, from experiments carried out in the light and dark in *D. melanogaster*. (From Bastock 1956.)

of antennae the mating success for both types of male was low. The winged males are presumably unable to stimulate antenna-less females as well as they can normal females: hence, the wings of males and antennae of females are important in courtship. Visual stimuli seem unimportant in the courtship of *D. melanogaster* because there is little difference in results between experiments carried out in the light or dark (Fig. 3.1). Ewing (1964) has demonstrated the importance of male wing area in courtship behaviour, since he found a fairly direct relationship between wing area as modified by temperature during pre-imaginal development, selective breeding, amputation, and mating success. In Fig. 3.2 the relationship between mating success and wing area is given, where wing areas were modified by amputation. Ewing, in fact, considers that about 80 per cent of the sexual stimulation is normally

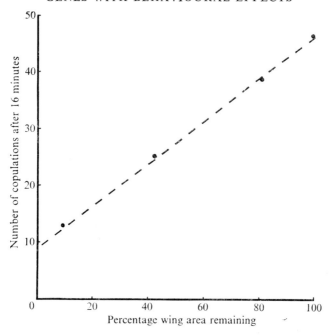

FIG. 3.2. Graph showing wing area plotted against number of copulations of *D. melanogaster* at sixteen minutes: the wing areas were varied by amputation. (After Ewing 1964.)

provided by wing display. The importance of wing-beat behaviour was shown by Williams and Reed (1944) in thirty-three isogenic mutant strains of *D. melanogaster*, where 24 per cent of the strains differed in wing-beat characteristics from wild-type normals. No doubt correlates with courtship behaviour occur in many cases. *Licking* is the final stage of courtship when the male goes behind the female, licks her genitalia with his proboscis, and attempts to copulate.

Bastock noted the activity of males every 1·5 seconds during the first 2·5 minutes of the courtship of wild-type and yellow males with wild-type females. The percentage of each courtship (disregarding periods when the male was not courting) that consisted of the three above-named elements, orientation, vibration, and licking, was assessed. From Table 3.2 it is clear that yellow males have a lower vibration percentage than do wild-type males. Furthermore, if the average bout length consisting of vibration plus licking is compared, yellow males have a significantly shorter bout length than do wild-type males. This is a reasonable measure of courtship activity, as licking is momentarily superimposed on vibration during a vibration bout. Thus yellow males are generally less vigorous than the wild-type males are in courtship behaviour. Furthermore, the orientation phase of yellow males is longer than that of wild-type males, showing that they take longer to become

TABLE 3.2

Analysis of courtship of wild and yellow males with wild females in D. melanogaster

	No. of records out of 100 in which courtship activities were shown	Courtship records			Bout length†	
		Orientation (O)	Vibration (V)	Licking (L)	Average (V+L)	(O)
Wild-type ♀ × wild-type ♂	92	72	22	6	3·9	5·5
Wild-type ♀ × yellow ♂	83	77	18	6	2·9	6·9
Probability	0·05	0·05	0·05	not significant	0·01	not significant

† Bout length is expressed in units of 1·5 seconds.

After Bastock 1956

motivated. Hence the yellow mutant alters behaviour by generally affecting sexual motivation and the quality of courtship stimulation of the males.

The idea that the behavioural change consists of the female's reaction to changed colour or scent of the males was rejected on several grounds. First, there was no difference in the behaviour of the wild female towards the two types of males during courtship. Second, when courting *D. simulans* females, the average bout length was greater in wild-type than in yellow males. Furthermore, *D. simulans* females rejected both types of male equally. Third, the relative success of wild and yellow males in the light and dark was unaltered; similarly, the relative success was unchanged in the comparison of females with and without antennae. If females reacted against the yellow males because of changed scent or appearance, it would be expected that the elimination of these stimuli would reduce the difference in the success of the two types of male.

The data discussed were collected after seven generations of crossing the wild stock with the yellow stock. In Table 3.1 (see page 11), data are given comparing the percentage success in one hour, before crossing the wild stock with yellow, and after seven generations of crossing. The former shows a significant difference between both sexes, and the latter between males only. Therefore, as well as the yellow locus being involved the genetic background is also relevant. It is unfortunate that there are rather few studies in which genetic background effects are discussed. It can probably be argued that under natural conditions (before back-crossing) the receptivity of yellow females would need to be relatively high in view of the low level of stimulus offered by yellow males (see Table 3.1, page 11).

Elens (1957, 1958) studied the mating behaviour of ebony and wild-type, and found that the gene ebony noticeably diminished the sexual activity of the males. This also depends on the cytoplasm; in other words, as in Bastock's work, variations in sexual activity are not entirely dependent on the major locus under study.

2. Choice experiments

Although there are a number of other reports in the literature showing variations between genotypes for mating behaviour yet to be reviewed, choice experiments will now be considered, where males of one or more genotypes are placed with females of one or more genotypes. Variations between different genotypes in choice experiments are presumably the consequences of courtship behaviour differences. An example comes from McKenzie and Parsons (1971), who reported on the mating behaviour of three lines which had been subjected to directional selection for 73 generations for high scutellar chaeta number by selecting, in each generation, ten flies of each sex with the highest chaeta number out of 100. The three lines, A, B, and C, had mean chaeta numbers of about 4, 6, and 16 respectively. Line C was also known to be homozygous for scabrous (*sca*), a second chromosome mutant. Flies aged 2–3 days were placed in various combinations in a mating chamber with an aspirator. As soon as a pair commenced mating, it was sucked into a trap and stored separately to await classification.

In the first experiment, fifteen flies of each sex of the three lines were placed in the mating chamber. The numbers mating in 15 and 60 minutes are given in Table 3.3. In both cases deviations from random mating occurred

TABLE 3.3

Numbers mating in 15 and 60 minutes in mating chambers containing 15 flies of each sex of lines A, B, and C in D. melanogaster *(based on 7 trials)*

Time		15 minutes				60 minutes		
♂	A	B	C	Total	A	B	C	Total
♀ A	0	0	0	0	2	1	1	4
B	1	0	0	1	8	8	2	18
C	32	12	2	46	45	34	4	83
Total	33	12	2	47	55	43	7	105
χ_2^2 for 1:1:1 ratios on marginal totals			♂ 31·96† ♀ 88·11†				35·66† 101·54†	

† $P<0·001$. After McKenzie and Parsons 1971

such that C females mated much more frequently than either A or B females, while the converse was true of C males relative to males from the other two lines. Therefore the mating capabilities of the two sexes were not identical. This was followed by studying numbers mating, taking the lines in pairs, and again using fifteen flies per sex (Table 3.4), since in this way possible complex interactions between lines might be eliminated. Comparing lines A and B, the χ_1^2 values for the 1:1 ratios on the marginal totals show little difference in the mating propensities of the two sexes for either line. On the other hand, a comparison of lines A and C shows that the females of line C mated almost to the exclusion of line A, and conversely A males mated more frequently than C males. The significant χ_1^2 values for the 1:1 ratios on the marginal totals in both sexes confirm the high mating propensity of C females compared with A females, and that C males have a lower mating propensity than A males. For the comparison between lines B and C, results almost parallel to the A, C comparison were obtained. These conclusions were confirmed by simple male- and female-choice experiments (that is, one male with two

TABLE 3.4

Numbers of D. melanogaster *mating in 15 and 60 minutes in mating chambers containing 15 flies of each sex taking the 3 lines A, B, and C in pairs*

A and B (based on 9 trials)

Time		15 minutes			60 minutes		
♂		A	B	Total	A	B	Total
♀	A	3	3	6	12	12	24
	B	4	4	8	17	18	35
Total		7	7	14	29	30	59

A and C (based on 8 trials)

♂		A	C	Total	A	C	Total
♀	A	0	0	0	3	1	4
	C	40	9	49	74	22	96
Total		40	9	49	77	23	100

B and C (based on 8 trials)

♂		B	C	Total	B	C	Total
♀	B	4	1	5	17	3	20
	C	42	9	51	68	26	94
Total		46	10	56	85	29	114

χ_1^2 for 1 : 1 ratios on marginal totals (which assess the relative mating propensities of the lines)

A, B	♀	0·29	2·05
	♂	0·00	0·02
A, C	♀	49·00†	84·64†
	♂	19·61†	29·16†
B, C	♀	37·79†	48·04†
	♂	23·14†	27·51†

† $P<0.001$.   After McKenzie and Parsons 1971

female types and vice versa). Thus the mating propensity of both sexes of line C is different from A and B, whereby there is an increase in the female, and a decrease in the male, mating propensities of C as compared with A and B.

The question then arises as to whether, when line C males and females are placed together alone, the numbers mating would be equivalent to those found by placing both sexes of lines A and B alone, since this would show that in C thresholds have been altered in both sexes relative to A and B. Such data are given in Table 3.5 for thirty flies of each sex for each line, in

TABLE 3.5

Numbers of D. melanogaster *mating in 15 and 60 minutes in mating chambers containing 30 flies of each sex for each of the lines, A, B, and C and* sca sca *(based on 5 trials)*

Line	15 minutes	60 minutes
A	19	37
B	18	38
C	30	52
sca sca	24	45

After McKenzie and Parsons 1971

turn, in the mating chamber. There are no significant differences between the numbers mating for any of the lines at either time interval. Thus, when mating is considered *within* a given line, without the presence of other lines, the numbers mating are relatively equivalent showing an over-all balance in line C, even though the thresholds in single sexes differ in comparison with lines A and B.

Because line C was homozygous for scabrous, its mating behaviour was compared with a laboratory stock homozygous for *sca*, but without the high chaeta number of C. Table 3.6 shows *sca sca* flies to be analogous to line C, since the mating propensities of *sca sca* and line C were essentially equivalent after 60 minutes, although at 15 minutes the *sca sca* males and C females were somewhat more successful than their respective counterparts. Furthermore, *sca sca* showed non-random mating with lines A and B in an analogous (but perhaps a little less extreme) way to line C. Therefore *sca sca* flies have mating propensities similar to, but somewhat less extreme than C, so that the behavioural differences between C on one hand, and A and B on the other, probably reside mainly in the region of the *sca* locus.

These results have analogies with Bastock's (1956) work on the yellow mutant, since the yellow male is less able to stimulate the wild-type female than the yellow female to the threshold required for mating to occur. As an example are some data of Parsons (1967*b*) in a mating chamber of a similar

TABLE 3.6

Numbers of D. melanogaster *mating in 15 and 60 minutes in mating chambers containing 15 flies of each sex for lines A, B, and C with homozygous* sca sca

A and *sca sca* (based on 5 trials)

Time		15 minutes			60 minutes		
♂		A	*sca sca*	Total	A	*sca sca*	Total
♀ A		7	3	10	10	3	13
sca sca		14	4	18	24	12	36
Total		21	7	28	34	15	49

B and *sca sca* (based on 5 trials)

♂		B	*sca sca*	Total	B	*sca sca*	Total
♀ B		4	5	9	8	6	14
sca sca		12	3	15	32	10	42
Total		16	8	24	40	16	56

C and *sca sca* (based on 5 trials)

♂		C	*sca sca*	Total	C	*sca sca*	Total
♀ C		7	14	21	14	21	35
sca sca		3	7	10	11	12	23
Total		10	21	31	25	33	58

χ_1^2 for 1 : 1 ratios in marginal totals (which assess the relative mating propensities of the lines to homozygous *sca sca*)

A, *sca sca*	♀	2·29	10·80†	
	♂	7·00†	7·37†	
B, *sca sca*	♀	1·50	14·00§	
	♂	2·67	10·29†	
C, *sca sca*	♀	3·90‡	2·48	
	♂	3·90‡	1·10	

† $P<0\cdot01$, ‡ $P<0\cdot05$, § $P<0\cdot001$.

After McKenzie and Parsons 1971

type to that used in collecting the above results, where yellow and wild-type (Canton-S) flies were mixed in equal proportions (twenty per genotype per sex). For five experiments, it was found that there were $26 +♀ × +♂$, $26 y♀ × +♂$, $5 +♀ × y♂$ and $18\ y♀ × y♂$ matings at the stage when three-quarters of the total possible matings had occurred, thus showing extreme non-random mating due to the lack of success of y males with $+$ females, as would be expected from Bastock's (1956) analysis already presented (which was not based on a choice situation).

Other mating propensity tests have been carried out and Spiess (1970) presents a good summary. It is difficult to review these experiments critically because of variations in technique, but they show conclusively that genetic differences between strains do have a considerable effect on mating behaviour and propensity. The main experimental designs used are:

(1) Two experiments consisting of females of types 1 and 2, with males of type 1 in one experiment, and type 2 in the other, thus:

$$(1+2)♀♀ × 1♂♂$$
$$(1+2)♀♀ × 2♂♂$$

This is a 'male-choice' experiment since the males can choose one of two types of female, although naturally the female type may be relevant in the choice, perhaps by differing in levels of receptivity to the males.

(2) Two experiments consisting of males of types 1 and 2, with females of type 1 in one experiment, and type 2 in the other, thus:

$$1♀♀ × (1+2)♂♂$$
$$2♀♀ × (1+2)♂♂$$

This is a 'female-choice' experiment, since the female can choose one of two males, although naturally the male type may be relevant in the choice; for example, by competing with each other for females.

(3) An experiment where females and males of two types 1 and 2 are put together, thus:

$$(1+2)♀♀ × (1+2)♂♂$$

which is a situation perhaps most closely corresponding to natural conditions, and is called a 'multiple-choice' experiment.

(4) Separate or single-pair matings, thus:

$$1♀ × 1♂,\ 2♀ × 2♂,\ 1♀ × 2♂,\ \text{and}\ 2♀ × 1♂$$

If there is a higher frequency of matings in a given time for the first two mating types (homogamic matings), then isolation would be inferred. This approach is most successfully used for fairly distinct races between which there is a fairly strong tendency towards homogamic matings. It does not represent a choice situation, but is included here for completeness.

* * *

FIG. 3.3. Two views of the Elens-Wattiaux (1964) mating chamber, which is used for the direct observation of mating behaviour: a description is given in the text. (After Ehrman 1965, from photographs supplied by her.)

As well as different experimental designs, there are various methods of scoring: for example, by direct observation of the matings either in the mating chamber or by removal from it, as in the experiments of McKenzie and Parsons (1971), or by dissection or progeny tests of females to seek evidence of sperm transfer. In many cases, whatever the method of scoring, it is necessary to mark certain types (see, for example, Southwood 1966). Several methods have been used, such as clipping the distal margins of the wings, marking with plastic colour dissolved in acetone, and dusting flies with a mixture of dye powder plus wheat flour. The critical point to ascertain is that the marking procedure does not affect mating. This can be verified by carrying out a control experiment.

With these comments we can look once again at the various experimental designs. The first design (male choice) presents no difficulties, since direct observation, or dissection, or progeny tests of the females enables a classification to be made. Multiple matings are likely for this design, and if the sequence for a given male is desired direct observation is necessary, although to obtain this information the mating chamber must contain very few males so that they can be followed individually. This particular design was used extensively in the earlier literature.

The second design (female choice) is simple if direct observation is possible. If not, scoring must be done by progeny testing which in many situations may not reveal the identity of the male, especially when closely related races from natural populations are tested.

The third design (multiple choice) is usually carried out with direct observation of the mating chamber, otherwise progeny testing of females is necessary. Direct observation is possible in a mating chamber of the type designed by Elens and Wattiaux (1964) (Fig. 3.3). This particular mating chamber is essentially a glass and wood sandwich. In the middle of the piece of wood, a circle is cut out to form the mating chamber; the bottom of the chamber is formed by a chequered canvas. Through the small lateral hole in the wood, a number of flies are introduced without anaesthesia. Any marking of flies will have been done at least a day previously to avoid effects due to previous anaesthesia. Quite a large number of flies can be introduced, say, up to sixty or more virgin pairs, but this will depend on the species. Copulating pairs do not generally move and so can be localized on the chequered canvas. This technique permits not only the observation of the types of males and females in a mating, but also the time at which it takes place, its sequence among other matings, and the duration of copulation; it does not permit the successive recording of matings of given single males unless very few flies are used. This method has been used often in recent years. Certain other designs for mating chambers have been devised. If a large number of flies is to be studied, and measurements are required on the copulating pairs of flies, then a mating chamber large enough for mating pairs to be extracted with an aspirator is necessary, but multiple matings for a given fly are then impossible. The data for the A, B, and C lines discussed above were collected using this technique. A modification used was to put flies of both sexes of the three lines together in one series of experiments, and other modifications are clearly possible.

The scoring procedure for the fourth experimental design (single-pair matings) is by direct observation of mating speeds (the time between placing the male and female together and the commencement of copulation), or if mating tends to be slow, as is likely if the strains are suspected of being rather distinct ethologically, female dissection or looking for evidence of progeny may be necessary.

3. Choice indices

From such experiments, variations in mating success can be quantified using various indices (see Parsons 1967a), which assess two main things: first, variations in the relative mating propensities of one or both sexes between genotypes, that is, variations in sexual vigour, which therefore give estimates of *sexual selection*, being the advantage certain individuals have over others of the same sex and species solely in respect to reproduction; and second, *sexual isolation*, which comes from comparing the relative proportions of homogamic and heterogamic matings. If mating is at random there is an expectation of equal numbers of homogamic and heterogamic matings, and sexual isolation occurs if there is an excess of homogamic matings.

Some of these indices may be illustrated, beginning with the male-choice situation (design 1). Thus, let there be n_1 females of type 1 and n_2 of type 2, together with males of type 1. Further, let $x_{1,1}$ and $x_{1,2}$ be the numbers of types 1 and 2 females inseminated respectively, and let $p_{1,1}=\frac{x_{1,1}}{n_1}$ and $p_{1,2}=\frac{x_{1,2}}{n_2}$, so that $p_{1,1}$ and $p_{1,2}$ represent the proportions inseminated. Stalker (1942) introduced an *isolation index*

$$b_{1,2}=\frac{p_{1,1}-p_{1,2}}{p_{1,1}+p_{1,2}}.$$

The index can range from $+1$ for 100 per cent homogamic matings to -1 for 100 per cent heterogamic matings. When the index is zero there is no isolation. To see if deviations from $b_{1,2}=0$ are significant, simple χ^2 tests can be carried out on the raw data for an expectation of $p_{1,1}=p_{1,2}$.

If the male is of type 2 and the two types of female are as before, a reciprocal isolation index

$$b_{2,1}=\frac{p_{2,2}-p_{2,1}}{p_{2,2}+p_{2,1}}.$$

can be calculated, where $p_{2,2}$ and $p_{2,1}$ are defined in an analogous way to $p_{1,1}$ and $p_{1,2}$.

These indices depend on the over-all proportion of females inseminated, that is, on the duration of the experiment, so that for comparisons of indices from different experiments there should be a similar over-all proportion inseminated. For this reason a number of experimenters terminate their experiments, so far as is possible, when about 50 per cent of females have been inseminated (for example, Ehrman 1964).

Joint isolation indices based on the combination of the pairs of experiments with males of types 1 and 2 respectively have been proposed. If there are equal numbers of females or couples of each of the two types, the *joint isolation index* comes to

$$\frac{x_{1,1}+x_{2,2}-x_{1,2}-x_{2,1}}{N},$$

where $N=x_{1,1}+x_{2,2}+x_{1,2}+x_{2,1}$, the total number of matings (Malogolowkin-Cohen, Simmons, and Levene 1965). If there are not equal numbers of females or couples, the arithmetic mean of the two indices $b_{1,2}$ and $b_{2,1}$ thus

$$b_{1,2}=\frac{b_{1,2}+b_{2,1}}{2}$$

is used.

Analogous indices can be calculated from female-choice data (design 2). From a multiple-choice experiment (design 3), all the indices described so far can be computed, although the biological situation in the multiple-choice situation differs from that in designs 1 and 2.

Turning to sexual selection indices, Bateman (1949) proposed an index which measures the relative mating propensity of females and is defined as

$$a_{1,2}=\frac{b_{1,2}-b_{2,1}}{2},$$

which is positive if there is an excess of matings of females of type 1, and negative if there is an excess of matings of females of type 2 in a male-choice experiment; a similar index can be derived from the female-choice experiment. Various indices have been proposed by other authors (see Levene 1949, Petit 1958, Parsons 1967a), but the ones presented probably suffice for most data.

Most of the indices presented have been given standard errors in the original papers so that their significance can be assessed. However, provided equal numbers of each sex are used, equal numbers of homogamic and heterogamic matings will be expected in the isolation indices if mating is at random, which can be tested with simple χ^2 tests. Even in more complex situations, observed data can be compared with what is expected assuming random mating, and the deviation from this expectation can be tested using χ^2 tests. In the case of coefficients measuring differences in the sexual vigour of two types of male or female, the expectations will be based on the proportions of the two types in the population initially.

4. Single genes and mating behaviour

As an example, Table 3.7 shows early data based on direct observation, using female- and male-choice experiments (Sturtevant 1915) for white-

TABLE 3.7

Results from male- and female-choice experiments between white-eyed and wild-type D. melanogaster

Male choice	Number of females mated	
Wild-type male	Wild-type 54	white-eyed 82
White-eyed male	Wild-type 40	white-eyed 93
Female choice	Number of males mated	
Wild-type female	Wild-type 53	white-eyed 14
White-eyed female	Wild-type 62	white-eyed 19

After Sturtevant 1915

eyed and a wild-type strain in *D. melanogaster*. The joint isolation index came to 0·097 in the male-choice experiment and to 0·026 in the female-choice experiment; thus, there is little evidence for choice of mates leading to sexual isolation as these values are close to zero. However, the relative mating propensity of wild-type females compared with white came to −0·303 in the male-choice experiments, while the relative mating propensity of wild-type males compared with white in the female-choice experiments came to 0·558; thus, there is some evidence for non-random mating due to differences in the vigour of sexual behaviour—that is, there is sexual selection. Similar indices can be computed for the data of McKenzie and Parsons (1971), although the experimental design is different. In any case, Tables 3.3, 3.4, and 3.6 give χ_1^2 tests on 1:1 ratios expected for equal mating propensities of given sexes.

A rather more complex example comes from Merrell (1949a), who studied the effect on mating of four sex-linked mutants in *D. melanogaster*, singly and in combination in the same fly, using male- and female-choice experiments. By using females heterozygous for the mutant—for example, raspberry (*ras*)—female-choice experiments could be carried out by progeny testing.

Thus for *ras* and + males mated with a *ras*/+ female, progeny are obtained as follows:

$$\frac{ras}{+} ♀ \times ras\ ♂ \rightarrow \tfrac{1}{2}\frac{ras}{ras} ♀ : \tfrac{1}{2}\frac{ras}{+} ♀$$

$$\tfrac{1}{2}\ ras\ ♂ : \tfrac{1}{2} + ♂$$

$$\frac{ras}{+} ♀ \times +\ ♂ \rightarrow \tfrac{1}{2}\frac{ras}{+} ♀ : \tfrac{1}{2}\frac{+}{+} ♀$$

$$\tfrac{1}{2}\ ras\ ♂ : \tfrac{1}{2} + ♂$$

The presence of mutant female progeny indicates the success of the mutant male.

Deviations from random mating occurred more often and were greater in the female-choice experiments. Thus wild-type males were much more successful than yellow (y) males (as expected—see §3.1), and were moderately more successful than cut (*ct*) and *ras* males, while forked (*f*) males were equivalent to the wild type in mating success. In combination, the effects of the mutant genes were mainly additive but in some cases involving *ras*, the additivity broke down; in particular, *ct ras* males were superior to either *ct* or *ras* males. In the male-choice experiments, deviations from random mating were most evident for the less vigorous males. Presumably the less vigorous male was rejected by the less receptive female so leading to deviations from random mating. In the case of vigorous males, the female's reaction to the male's courtship pattern is probably less important than for less vigorous males. The conclusion was that selective mating was important in these experiments, but that there was no tendency for like phenotypes to mate leading to an excess of homogamic matings indicating sexual isolation.

In subsequent experiments, Merrell (1953) studied gene frequency changes in populations each containing one of the four sex-linked genes. Initially the populations had equal numbers of mutant and wild-type males and only heterozygous females, so that the gene frequency of the mutants was 0·5. In all cases the frequency of the mutant genes fell quite rapidly over a number of generations. The experimental results agreed with predictions based on taking into account levels of selective mating for these mutants, implying that selective mating was a major factor in the decline in frequency of the mutant genes. This provides a demonstration of the importance of selective mating as a component of fitness in the populations under study.

The various experiments surveyed here with mutant genes, and those surveyed by Spiess (1970), show the existence of sexual selection determined by differences in vigour of one or both sexes. Spiess concluded from his survey that mating success depends far more on the female's control, by her acceptance of particular mates, than on the male's choice of a mate, or—put succinctly—the males provide the diversity from which the females do the selecting. While this may be true over a long period of time, it could well be that the importance of one sex or the other may depend on the *time* after the commencement of the experiment at which results are assessed (see Chapters 4 and 5).

A difficulty in interpretation comes at the level of technique, in that there are several types of mating chamber, as well as several main experimental techniques (male- , female- and multiple-choice, and pair-mating). Also, in some cases pairs are removed with an aspirator as mating occurs and in other cases this is not done. Even so, it seems that a given genotype yields similar results for differing techniques, although this has not been greatly tested. Barker (1962*a*) found somewhat varying results between methods, using yellow and wild-type strains. He found that male choice and pair matings indicated less isolation between mutant and wild-type phenotypes than do female-choice and multiple-choice matings. The degree of selective mating was significantly affected by mating period, which confirms the need in sexual isolation tests to analyse the data at the stage when a fixed proportion of flies have mated, say 50 per cent, so as to obtain comparable isolation indices. He also reported that the number of flies per mating bottle is an environmental factor that must be considered.

This leads on to environmental factors, which are of critical importance. For example, *D. melanogaster* and *D. persimilis* show positive, and *D. pseudoobscura* negative, phototaxis (Pittendrigh 1958). In *D. pseudoobscura* strong illumination leads after a few minutes to cessation of movement in the chamber (Spiess and Langer 1964*a*). Furthermore, responses to light vary according to other environmental factors such as temperature or water balance (Pittendrigh 1958). An illustration of the level of complexity which is possible is Lewontin's (1959) observation that *D. pseudoobscura* is negatively phototactic under conditions of low excitement, but that if the flies are forced to walk rapidly or fly, they lose their negative phototaxis and become strongly attracted to light. Two species, *D. subobscura* (Philip, Rendel, Spurway, and Haldane 1944, Rendel 1945) and *D. auraria* (Spieth and Hsu 1950) require light in order to mate, and a number of species give differing insemination frequencies in the light and the dark, while a third group of species is independent of light (Grossfield 1966). In the light-dependent species, *D. palustris*, it was found that the ability to mate in darkness is probably inherited as an autosomal recessive (Grossfield 1966).The importance of wings in *D. subobscura* and *D. auraria* as visual cues is discussed by Grossfield (1968). It is also of interest that *D. melanogaster* can mate freely in the dark, but a sibling species, *D. simulans*, is greatly inhibited by the dark (Spieth and Hsu 1950).

Within the species *D. melanogaster*, Elens (1958) has shown that the sexual activity of males of ebony and wild-type flies, and the heterozygotes between them, depended on the temperature at which the experiments were carried out. Rendel (1951) reported that wild-type and vestigial males mate more readily in the light than in the dark, which he attributed to increased general activity in the light; conversely, ebony males were more successful in the dark. Jacobs (1961) studied laboratory populations of *D. melanogaster*, begun with an initial high frequency of ebony flies and a low frequency of non-ebony, and found that the frequency of ebony flies fell more rapidly in the light than in the dark in agreement with Rendel's results. Multiple choice studies in which *ee* competed with ++ showed ++ to have a mating advantage in the light, while *ee* had a mating advantage in the dark. On the other hand, +*e* males showed a mating advantage over ++ in the light, and still more in the dark. Thus environmental factors that affect mating behaviour (see discussion in later chapters), coupled with the variety of techniques employed, present major difficulties when comparing experiments carried out by different workers. At this point it is desirable to present an exact description of techniques in published work so that differences between experimenters are made obvious.

Geer and Green (1962) investigated the mating behaviour of *D. melanogaster* as assessed by competition between males with different white-eye colour alleles in a uniform genetic background. It was found that in an alternating light–dark environment males possessing the more pigmented eye colour were more successful in mating, but in the dark mating was at random. The comparison was between white-eyed flies (w) and those carrying various white-apricot (w^a) alleles. No effect of the female's genotype or phenotype on mate selection was noted. Therefore, these experiments show the importance of selective mating associated with visual pigmentation and, furthermore, that visual stimulus is of importance for mate selection by males. Precisely how these stimuli function in the mating behaviour of males is not clear, but for some species of *Drosophila* which fail to mate in the dark, such as *D. subobscura* and *D. auraria*, this factor is all-important (although it must be noted that Bastock (1956) found no evidence for visual stimuli).

Normally *D. melanogaster* which are homozygous *v;bw* (vermilion brown) have 'pale-sherry-coloured' eyes, with a marked attenuation of visual acuity in terms of optometer response. The mutation at the vermilion locus results in a block in brown pigment synthesis; thus, vermilion flies have bright-red eyes. However, the vermilion locus block can be by-passed by supplying kynurenine in the larval diet, which leads to the formation of the brown pigment in the eye. The effect of kynurenine in enhancing the male mating success of *v;bw* flies is shown in Table 3.8 compared with those not so treated (Connolly, Burnet, and Sewell 1969). The results provide good evidence in support of the hypothesis that the mating disadvantage shown by flies lacking eye pigments is due to the sensory defect accompanying the absence of screening pigment in the compound eye, and that this can be alleviated by

TABLE 3.8

Results of competition in D. melanogaster *between kynurenine-treated* v;bw *males and* bw *males, and between kynurenine-treated* v;bw *males and* v;bw *males*

Total number of mating competitions	Description of male	Number of males mating	χ_1^2 for 1:1 ratio
126	bw	52	3·5
	Kynurenine-treated v;bw	74	
83	v;bw	15	30·12†
	Kynurenine-treated v;bw	68	

† P<0·001. After Connolly, Burnet, and Sewell 1969

biochemical supplements. Connolly *et al.* (1969) therefore suggest that the role played by vision in the courtship behaviour of *D. melanogaster* has been underestimated. Analysis of courtship behaviour of males with pigmented and non-pigmented eyes indicated that the inferior courtship of *v;bw* males results in difficulties in establishing and maintaining contacts with females. Flies, *v;bw*, with attenuated visual acuity were found to have a significantly shorter bout length of wing vibration in males, which has analogies with Bastock's (1956) data on yellow flies. More generally, mating tests with flies possessing varying amounts of brown pigment revealed a close correlation between mating success and eye-pigment density. On the other hand the absence of red pigment does not, as in *bw* flies, lead to any attenuation of optometer response (Burnet, Connolly, and Beck, 1968), neither does it seem to affect courtship latency or duration as occurs in *v;bw* flies (Connolly *et al.* 1969). An earlier observation shows the general correlation between brown eye-pigment density and fitness (Parsons and Green 1959) using competition experiments. Where the *v* gene is suppressed genetically by a suppressor gene, the fitness of the flies concerned is increased under conditions of high competition. The effect of the suppressor gene on competition can be imitated by administering kynurenine which similarly permits brown pigment formation. In other words by-passing the block caused by the vermilion gene affects both mating success and competitive ability, and so possibly fitness in general.

Other genes have been shown to be involved in mating behaviour (see reviews of Manning 1966, 1968*a*, Thiessen, Owen, and Whitsett 1970). The only conclusion is that, with the exception of forked *f*, which is neutral so far as mating success is concerned (Merrell 1949*a*), all other mutants show a depression of mating success. Apart from those genes already mentioned, these others include the sex-linked gene Bar *B*, and the autosomal gene black b. Of the numerous loci available in *Drosophila*, the number of studies is therefore small and much more work would be of interest. Unfortunately, with the exception of Bastock's (1956) work on yellow y, no studies of the effect of the background genotype have been carried out. Although tedious to take into account, it is to be hoped that this aspect may be investigated in greater detail in future: the same point applies to other types of behaviour which form the subject of the next section.

5. Other forms of behaviour

So far mutants which are mainly involved with sexual behaviour have been considered; however, several other mutants have recently been shown to have other behavioural effects. Thus Kaplan and Trout (1969) described four neurological mutants in *D. melanogaster* produced by feeding adult Canton-S males on a 0·25-M solution of ethyl methane sulphonate (EMS). Four sex-linked behavioural mutants were found, all having rapid leg-shaking following etherization. Although this phenotype is common to all mutants, other aspects of anaesthetized and conscious behaviour distinguish them from each other and from wild flies. The four genes involved are Hyperkinetic (alleles Hp^{1P} and Hp^{2T}), Shaker^5 (Sh^5), and Ether à go-go (Eag). Just as many years of study of fly morphology have revealed numerous mutants, the same is likely to be true when behavioural variants are studied in detail.

Benzer (1967) described a procedure whereby *Drosophila* populations can be fractionated according to their behavioural responses on repeated trials for phototaxis. Various genotypes were found to show great differences, and, by application of the method to the progeny of flies treated with EMS, sex-linked behavioural mutants showing a loss of phototaxis were isolated in one generation. While this technique seems capable of general application, it is difficult to carry out, which perhaps explains why rather scanty additional data have resulted from its use. Earlier work has shown that certain eye-colour mutants have an effect on phototropism; thus, brown-eyed flies were attracted to a light source more rapidly than white-eyed flies (Scott 1943). Fingerman (1952) studied phototaxis to monochromatic light for a number of *D. melanogaster* mutants which differed in their amounts of eye pigmentation—namely, wild-type+, white w, white-apricot w^a, vermilion v, brown bw, and sepia se—and he found that in a choice between illuminated and non-illuminated arms in a Y-maze, there was greater phototaxis in flies with more heavily pigmented eyes. Médioni (1959, 1963) found that yellow flies of *D. melanogaster* were more strongly phototactic than wild-type, and that Bar were less so. The Bar eye result is perhaps to be expected because of the lower number of facets in the eye. The strong phototactic behaviour of yellow suggests the possibility of a 'dermal' light sense in *Drosophila*, since one effect of the yellow gene is to decrease the tanning of the cuticle.

For optometer response, Kalmus (1943) obtained results paralleling those just described. Thus, optometer response was reduced in *D. melanogaster* in Bar-eyed flies, the decrease in response paralleling the reduction in facet number. The response was thus best in wild-type flies, and progressively worse in the order *B*+ females, *B* males, and *BB* females. White-eyed-flies, either genetically *w* or homozygous *bw;st*, had no optometer response, and w^a flies were markedly worse than wild-type flies as were *v;bw* flies which both have a light eye colour (see §3.4 for further evidence).

Therefore, there are rather few reports of single genes associated with forms of behaviour other than sexual behaviour variations. This is understandable since all other forms of behaviour need rather more complex apparatus for their assessment, and often variations between apparatus supposedly measuring the same form of behaviour can lead to complications. However, it can be anticipated that searches will be made in the future for non-sexual behavioural mutants, either natural or induced, as well as for non-sexual behavioural correlations of mutants with effects on sexual behaviour (and vice versa). In any case, Chapter 6 shows that a number of traits such as geotaxis, phototaxis, locomotor activity, optometer response, and chemotaxis respond to directional selection, demonstrating the existence of genes controlling these traits segregating in populations.

4

INVERSION SEQUENCES AND MATING BEHAVIOUR

1. Associations with karyotypes

IN natural populations of *D. pseudoobscura*, *D. persimilis*, and many other species of *Drosophila*, polymorphisms for chromosomal inversion sequences are common. The inversion sequences are effectively single gene complexes, since few recombinants are viable from the heterokaryotypes due to the mechanics of meiosis. Experimental work in *D. pseudoobscura* has shown that the heterokaryotype is often at a selective advantage over the corresponding homokaryotypes. Thus, the introduction of two sequences, Standard (*ST*) and Chiricahua (*CH*), into population cages at 25°C leads to ultimate equilibrium frequencies of about 0·7 *ST* and 0·3 *CH*, irrespective of the initial inversion frequencies (Wright and Dobzhansky 1946). The simplest explanation of the equilibrium is that the heterokaryotypes are at a selective advantage over the homokaryotypes, since this can be shown theoretically to be a condition for a stable equilibrium (Fisher 1922, and see appendix to this chapter). Such matters are further considered from the experimental point of view in Chapters 12–14 with certain theoretical complications. In this chapter, observations on variations in mating behaviour for carriers of karyotypes derived from natural populations are discussed, although it will become abundantly clear that the situation is much more complex than that presented in the appendix.

In *D. pseudoobscura* Spiess and Langer (1964a) have shown substantial mating speed differences between homokaryotypes derived from stocks collected at Mather, California. They used homokaryotypes for the *ST*, *CH*, Tree Line (*TL*), Pikes Peak (*PP*), and Arrowhead (*AR*) inversions. Mating was carried out at 25°C for flies 6 days old, using ten pairs per mating chamber. Fig. 4.1 shows the cumulative percentage curves for homokaryotypic homogamic matings over a one-hour period. The only contrasts not significant at 60 minutes are *AR* versus *ST*, and *CH* versus *TL*. The homokaryotypes therefore appear to have gene complexes controlling mating speed. It is interesting that at Mather the observed frequencies of the karyotypes were *ST*=40·1 per cent, *AR*=36·0 per cent, *CH*=10·8 per cent, *TL*= 7·2 per cent, and *PP*=4·3 per cent which approximately parallels the relative mating speeds. It is tempting to suppose that mating speed is a major factor in maintaining the observed frequencies of the chromosomes in this population

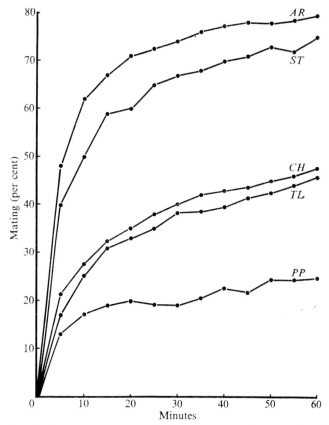

FIG. 4.1. Cumulative percentage curves for homokaryotype homogamic matings over a one-hour period (based on 300 pairs per curve) in *D. pseudoobscura*. AR = Arrowhead, ST = Standard CH = Chiricahua, TL = Tree Line, PP = Pikes Peak (From Spiess and Langer 1964*a*.)

(Spiess and Langer 1964*a*), but this interpretation may be naïve in view of the number of components of fitness known. In any case, whether this is true or not, it is clear that differences in mating speeds may be important in controlling the frequencies of the chromosomes in natural populations.

On the grounds that the heterokaryotypes contribute as much or more to future generations as the homokaryotypes, Kaul and Parsons (1965) studied the average mating speed and the duration of copulation of all possible combinations between the three karyotypic combinations ST/ST, ST/CH, and CH/CH, using vials with single pairs of flies, so making it possible to assess the duration of copulation. In Table 4.1 the mean numbers mating out fifty in 5 minutes are given. Inspection of the marginal frequencies shows that males of the ST/ST and ST/CH karyotypes have similar mating

TABLE 4.1

Mean numbers mating out of 50 in 5 minutes and mean durations of copulation in D. pseudoobscura

Mean numbers mating out of 50 in 5 minutes

Karyotype of male		ST/ST	ST/CH	CH/CH	
Karyotype of female	ST/ST	39·5	37·5	26·5	34·5
	ST/CH	36	40	23·5	33·17
	CH/CH	37	33	26·5	32·17
		37·5	36·83	25·5	33·28

Analysis of variance

Source of variation	d.f.	F	P
Female karyotypes	2	0·36	
Male karyotypes	2	12·84	<0·01
Replicates (crowding levels)	1	5·47	<0·05
Error	12		

Mean durations of copulation (minutes)

Karyotype of male		ST/ST	ST/CH	CH/CH	
Karyotype of female	ST/ST	5·08	4·22	3·17	4·16
	ST/CH	5·49	4·47	3·82	4·59
	CH/CH	5·95	4·38	3·55	4·63
		5·51	4·36	3·51	4·46

Analysis of variance

Source of variation	d.f.	F	P
Females	2	9·30	<0·001
Males	2	134·86	<0·001
Replicates (crowding levels)	1	3·15	=
Females × males	4	2·35	=
Females × replicates	2	1·09	=
Males × replicates	2	3·19	<0·05
Females × males × replicates	4	4·64	<0·001
Error	684		

After Kaul and Parsons 1965

frequencies, and males of CH/CH karyotype have a lower mating frequency. There is little difference between karyotypes for females; thus, mating frequencies are almost entirely male-determined, as is confirmed by an analysis of variance in Table 4.1. From this table it is clear that duration of copulation is mainly male-determined as confirmed by an analysis of variance. The shortest durations are for CH/CH males, followed by ST/CH and ST/ST. There is thus a negative correlation between mating frequency and duration of copulation for different karyotypes in the males. Hence the sums of the mean mating speed and duration of copulation approximate more closely to each other than do the component times (Table 4.2), so that mating speed

TABLE 4.2

Comparison of mean mating speeds and durations of copulation (minutes) in D. pseudoobscura

Karyotype	Females			Males		
	Mating speed	Duration of copulation	Sum	Mating speed	Duration of copulation	Sum
ST/ST	2·33	4·16	6·49	2·03	5·51	7·54
ST/CH	2·69	4·59	7·28	2·15	4·36	6·51
CH/CH	2·65	4·63	7·28	4·76	3·51	8·27

After Kaul and Parsons 1965

and duration of copulation may possibly be regarded as an integrated system controlled mainly by the male. The sum of the components is less for the heterokaryotypes than the homokaryotypes in males, and therefore represents an example of heterokaryotype advantage restricted to males, assuming that there is some selective advantage in completing mating and copulation rapidly. Presumably those individuals completing mating and copulation most rapidly would most readily leave genes in subsequent generations.

For duration of copulation in *D. melanogaster*, Merrell's (1949*b*) earlier results are in agreement with Kaul and Parsons (1965) in showing male determination based on experiments with two strains, Lausanne Special and Oregon-R. However, mating speed was determined more by females (but see Chapter 5 for a more detailed discussion of differences between strains). Spiess (1968*a*) has confirmed the male determination of duration of copulation in *D. pseudoobscura* for various karyotypes.

Turning to other observations on mating speed, Spiess, Langer, and Spiess (1966) studied mating speeds for a number of combinations using the technique of ten pairs of flies per mating chamber, as for data described earlier (see Fig. 4.1, page 32), and found that male heterokaryotypes consistently had a higher mating speed than the corresponding homokaryotypes, since in every case the value for heterokaryotypic males was superior to both of the corresponding homokaryotypes. In females there was no consistent superiority (that is, greater receptivity), so that the heterosis displayed is clearly due to the greater activity of the males, persistence in courtship, or to greater female acceptance of heterokaryotype males, probably because of the males' increased sexual activity.

The importance of males compared with females also emerges from Kaul and Parsons (1966), where two series of choice experiments were carried out consisting of one female with three males, and the reverse (three females with one male), to see whether competition between individuals of the same sex occurs. The mean period elapsing to the first mating was 0·53 minutes in the experiments with three females and 1·40 minutes in those with three males. The likely interpretation is that in the experiments with three males interference between males occurs, so delaying mating, whereas in the reverse situation with three females, one male tends to mate more rapidly having no competition from other males and little interference from females. Spiess and Ehrman (see Spiess 1970) have recently confirmed these observations with other karyotypes of *D. pseudoobscura*, but in *D. persimilis*, E. B. Spiess and L. D. Spiess (1969) did not find that an excess of males delayed the average time to mating (ratio of 8:2), although in 30 minutes of observation more matings occurred when females were in excess than when males were in excess. However, as discussed below, the females play a greater part in determining mating speed in *D. persimilis* than in *D. pseudoobscura*. E. B. Spiess and L. D. Spiess (1969) stress the point made by Parsons (1967a) that the assessment of experiments of this type may well depend on the time of observation.

In *D. persimilis*, Spiess and Langer (1964b) studied mating speeds of the Whitney (*WT*) and Klamath (*KL*) inversions. Direct observation was carried out on groups of ten flies of each sex at 20°C for flies aged 10 days. As well as homogamic matings between the *WT*/*WT* and *KL*/*KL* homokaryotypes, the two possible heterogamic matings were carried out. Both crosses with *WT*/*WT* females show greater mating speeds than crosses with *KL*/*KL* females; thus, the *WT*/*WT* females mate rapidly and the *KL*/*KL* females mate slowly, irrespective of the male karyotype. Hence, in contrast with the *D. pseudoobscura* results, the karyotype of the female seems relatively more important in determining mating speed. However, for a given female karyotype, *WT*/*WT* males have a somewhat faster mating speed than *KL*/*KL* males. Although the difference is relatively small compared with the difference between karyotypes of females, it shows that the karyotype of both sexes plays a part in determining mating speed. In experiments with heterokaryotypes *WT*/*KL*, Spiess and Langer (1964b) found them to have a mating speed intermediate between the homokaryotypes for homogamic matings, although if the heterokaryotypes were mated to homokaryotypes, the mating speed was fast or slow according to the mating speed of the homokaryotype.

Observations on actual mating behaviour have shown that WT/WT females accept males readily, while KL/KL females tend to refuse males, and that WT/WT males court more actively than do KL/KL males. The differences for the various combinations can therefore be interpreted in terms of the relative intensities of the copulation tendency of males and the acceptance (or conversely the avoidance) tendency of females. Earlier experiments (Spiess and Langer 1961) demonstrated for matings over a 24-hour period, that male mating propensity was the important feature as WT/WT males mated significantly more than KL/KL males, with no large differences according to the karyotype of the females. Thus, depending on the time after commencing the experiment, somewhat different results may occur, which are interpretable in terms of factors such as the relative intensities of copulation and acceptance tendencies, and the likelihood of multiple matings for a given individual which certainly occurs much more frequently for males than females. This temporal factor is discussed further in Chapter 5.

The results described so far are for flies derived from White Wolf, California. Spiess and Langer (1964b) studied mating speeds for homogamic matings between the two homokaryotypes from Mather, California, and found that WT/WT homogamic matings were somewhat more rapid than KL/KL homogamic matings, in agreement with the results just described, but the difference was far less marked than at White Wolf. This provides evidence for some differences between localities, without, however, changing the over-all picture.

Subsequently Spiess and Spiess (1967) compared the mating propensity of flies collected in Humboldt County, California (a region where freezing rarely occurs and the mean daily range is very narrow, namely, 14°–20°C in summer), with that of White Wolf flies found where freezing often occurs at night in summer (the daily average is about 15°C but the daily range may be from -5.5°C to $+26.4$°C in summer). The frequencies of WT, KL, and Mendocino (MD) inversions varied between localities, and the evidence suggested that the commoner chromosome arrangement has the higher mating propensity under most conditions, which is in agreement with the data in *D. pseudoobscura* already discussed (Spiess and Langer 1964a). The Humboldt strains were found to have better survival and 30-minute mating percentages at 25°C as compared with 15°C, while the high Sierra strains, of which White Wolf is one, found the higher temperature deleterious, which generally agrees with the environmental differences observed.

Among other species *D. pavani*, living in central Chile and on the eastern slopes of the Andes in Argentina, has been looked at in some detail. In most natural populations the heterokaryotypes exist in fairly uniform frequencies. Mating ability of the males was assessed by using virgin females of a sibling species *D. gaucha* having a standard gene arrangement. It was found that males heterozygous for the gene arrangements under study were superior in mating activity to the corresponding homokaryotypes within the same population (Brncic and Koref-Santibañez 1964). It is argued, therefore, that the superiority in mating activity of the heterokaryotypes may be an important factor in the maintenance of balanced polymorphisms in natural populations.

In *D. robusta*, Prakash (1968) has presented evidence of interactions between second and third chromosome karyotypes affecting mating speed. Such interactions might well be expected to be quite common, since the observed phenotype is a product of the whole genotype in a given environment, although frequently certain parts of the genome are more important than others. Although interactions might be expected to be common, evidence for between-chromosome interactions is restricted, which is no doubt partly due to the difficulty of setting up the appropriate breeding programmes for their detection.

2. Environmental factors

In conclusion, inversion karyotypes are associated with differences in mating behaviour. An extremely difficult problem is that of extrapolating from laboratory experiments carried out under controlled conditions to the wild. However, laboratory experiments should provide clues as to the types of behavioural phenomena to be studied in the wild, and the extent to which traits vary with the environment. It is therefore essential to define precisely the environmental conditions used, since probably almost any environmental factor may affect mating behaviour (see also Chapter 3). For example, repeating experiments on successive days may show the existence of substantial but often undefinable interactions. Nevertheless, certain environmental factors have been studied and are therefore definitely known to affect mating propensity. When cultured at cool temperatures (15°C) there is much less difference between the performances of *AR* and *PP* than at 25°C (Spiess *et al.* 1966). Also, in *D. persimilis*, *WT* and *KL* karyotypes differ in their optimal temperatures, with *WT* mating faster at cool temperatures and *KL* faster at warm temperatures than occurs in the reverse situation (Spiess 1970).

Larval density is another environmental factor affecting mating behaviour. Mating speeds in *D. persimilis* were substantially lower at a larval density of 500 larvae per vial compared with a lower larval density of fifty per vial (Spiess and Spiess 1969). In *D. pseudoobscura* and *D. persimilis* less sexually active forms were found to mate more when multiple pairs (greater than five) were in the chamber than when single and double pairs were present (Spiess 1970). Karyotypes with high mating propensities did not show this effect. The observations conform to a suggestion of Manning (1967) that receptive females require a certain total amount of courtship before mating, and that those which are slower to mate may simply need more contact from courting males.

Heterosis may be more extreme in 'stress' environments than in more optimal environments, as shown by Parsons and Kaul (1966) in *D. pseudoobscura* under conditions of heat stress, and by Spiess and Spiess (1967) in *D. persimilis* under conditions of cold stress (see Chapter 14 for further discussion). Heterosis for several fitness traits seems to be more pronounced under conditions of environmental stress than in optimal environments, whether the stress is high temperature, low temperature, or desiccation. A consequence is that heterokaryotypes may show less variability over a multiplicity of environments than homokaryotypes, as has been shown for a number of fitness traits including behavioural traits (Parsons and Kaul 1966; and for references, Parsons and McKenzie 1972).

Therefore, the need to pay attention to environmental factors is again stressed. For inversions segregating in the wild, this assumes more importance from the point of view of the role of mating behaviour as a component of fitness, than it does for the genes discussed in Chapter 3, which—although instructive to study—would normally be rare in natural populations. It is not possible to say more at this stage, but a perusal of some of the topics covered in the ecological genetics part of the book will reveal some rather complex variables (especially environmental stresses) which have been found to be of importance in determining the fitnesses of various genotypes, and which may interact with mating behaviour. The only valid way of assessing the importance of mating behaviour in natural populations is to study it in all the environments to which that population is likely to be exposed—clearly impossible at present.

APPENDIX

EQUILIBRIA FOR A LOCUS WITH TWO ALLELES A AND a HAVING DIFFERENT FITNESSES

Let the two alleles at a locus A and a have gene frequencies p and q, such that $p+q=1$. Under random mating, the gene frequencies will be fixed from generation to generation except for random variation. The simplest way of showing this is to follow what happens under random union of gametes, which in many cases is implied by random mating. The gametes in each sex can be represented formally as $pA+qa$, and if these unite at random the genotypic proportions in the next generation will be given by

$$(pA+qa)^2 = p^2 AA + 2pq\, Aa + q^2\, aa.$$

In this generation, the gene frequency of A is given by the frequency of AA plus half the frequency of $Aa = p^2 + pq = p(p+q) = p$, and similarly the gene frequency of $a = q$. This is the Hardy-Weinberg Law, so named after its two discoverers. This result is strictly true only in infinitely large populations, since in finite populations p and q will vary somewhat, due to random variation.

This assumes all genotypes to be equally fit, and we must now introduce the complication of different fitnesses for each genotype. Let the fitnesses of genotypes AA, Aa, and aa be $1-s$, 1, and $1-t$, respectively. The genotypic proportions before and after selection will be:

Genotypes	AA	Aa	aa	Total
Fitnesses	$1-s$	1	$1-t$	
Frequencies before selection	p^2	$2pq$	q^2	1
Frequencies after selection	$p^2(1-s)$	$2pq$	$q^2(1-t)$	$\overline{W}$

The average fitness of the population after selection is $\overline{W} = 1 - sp^2 - tq^2$. If p^1 and q^1 represent the gene frequencies of A and a in the next generation, then

$$p^1 = \frac{p^2 - p^2 s + pq}{\overline{W}} = \frac{p - sp^2}{\overline{W}}$$

and

$$q^1 = \frac{pq + q^2 - q^2 t}{\overline{W}} = \frac{q - tq^2}{\overline{W}}.$$

It is necessary to divide by $\overline{W}$ so that $p^1 + q^1 = p + q = 1$. At equilibrium, the gene frequencies are fixed from generation to generation, so that if the change in gene frequency p from generation to generation is written as Δp, it will be expected that

$$\Delta p = p^1 - p = 0$$

at an equilibrium, or

$$\Delta p = \frac{p - sp^2}{\overline{W}} - p = \frac{pq(tq - sp)}{\overline{W}},$$

which equals 0 when $p=0$, $q=0$, or $tq = sp$. The first two solutions are trivial when either the population will be AA or aa, which do not correspond to a polymorphic situation. From the last equation we find $p_e = \frac{t}{s+t}$ and $q_e = \frac{s}{s+t}$ as equilibrium values, which depend only on the selective values s and t. Therefore, irrespective of the initial values of p and q, the same equilibrium is expected.

For p_e and q_e to exist, either $s, t>0$ or $s, t<0$, since otherwise p_e or q_e would be <0 which is impossible. For these two conditions, we must examine the stability of the equilibrium. A *stable equilibrium* occurs if after there is a small displacement from p_e, as may occur by chance in a finite population, the population tends to return to p_e in subsequent generations. If p is slightly $>p_e$ in a sample for $s, t>0$, this means $sp>tq$ and hence $\Delta p<0$, so that p will decrease in the next generation and will thus approach p_e. Similarly, if p is slightly $<p_e$, then $sp<tq$, hence $\Delta p>0$ so that p once again approaches p_e. In both bases, therefore, selection tends to change the gene frequency towards the equilibrium value p_e, and so represents a stable equilibrium. An *unstable equilibrium* occurs if the displacement from p_e is accentuated generation by generation, and it can be shown, using an argument as above, that this occurs for $s, t<0$. Thus, in conclusion, a stable equilibrium may be expected if $s, t>0$, or $Aa>AA$, aa in fitness. For the inversions of *D. pseudoobscura*, fitnesses of *ST/ST*, *ST/CH*, and *CH/CH* of 0·7, 1, and 0·3, respectively, corresponding to $s=0·3$, and $t=0·7$ are realistic at 25°C, and these give $p_e=0·7$ and $q_e=0·3$ corresponding approximately to the frequencies observed in population cages at equilibrium. Thus heterokaryotype advantage provides a reasonable explanation of the results in the population cages.

However, it must be pointed out that the occurrence of a polymorphism does not always depend on the fitness of the heterozygote being greater than that of the homozygotes. A polymorphism is a method of maintaining genetic variability in an out-bred population, and any method of maintaining such variability, whether it be environmental heterogeneity, mating behaviour variations, non-random fertilization, or density-dependent fitnesses of genotypes, may in theory lead to a polymorphism. Some of these mechanisms are discussed in later chapters of the book.

5
QUANTITATIVE TRAITS

1. Phenotypic and genotypic variability.

MANY behavioural traits are quantitative rather than qualitative. This means that it is not possible to split the population under study into discrete groups for the trait. Variation of this type, without natural discontinuities, is referred to as continuous variation. The frequency distributions of many quantitative traits approximate more or less closely to the *normal distribution* (Fig. 5.1). The normal distribution can be completely described in terms of

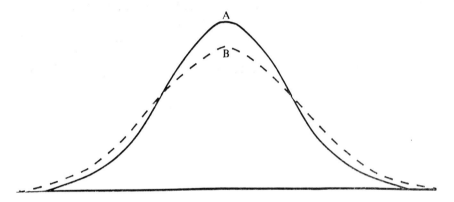

FIG. 5.1. Two normal distributions with the same mean and different variances (curve B has a larger variance than curve A).

two quantities or *parameters*. One is the mean or average value. If x_i is an individual observation and there are n observations, then the mean x is given by

$$\bar{x} = \frac{\Sigma x_i}{n}.$$

The other quantity is an expression of variability around the mean. In some cases the variability around the mean will be small, and in other cases large (see Fig. 5.1). The term for the quantity measuring variability is the *variance*, which is estimated as

$$\frac{1}{n-1} \Sigma(x_i - \bar{x})^2.$$

The square root of the variance is the *standard deviation*. It can be shown that 95 per cent of the population of a normally distributed trait lies within two standard deviations of the mean. If it is impossible to assume a normal distribution, it may be possible to find a suitable algebraic transformation which will convert the data to an approximate normal distribution. Much of the theory of quantitative genetics is based on the assumption of a normal distribution.

Assuming that a continuously varying trait is partly under genetic control, it must be asked how the intrinsically discontinuous variation caused by genetic segregation is converted to the continuous variation of quantitative traits. Suppose two individuals A/a. B/b are crossed together, where A,a and B,b are gene pairs at two unlinked loci, and further suppose that genes A and B act to increase the measurement of a quantitative trait by one unit, and genes a and b act to decrease the trait by one unit. It is perhaps less confusing to write $A/a.B/b$ as $+/-.+/-$, counting A and B genes as $+$genes, and a and b as $-$genes. Counting the number of $+$ and $-$genes will give a metrical or quantitative value for a genotype.

The above cross will give five genotypes distributed as in Fig. 5.2. The most frequent genotype is $+/-.+/-$, having a genotypic value of 0, which is the mean genotypic value, and the least frequent genotypes are the two

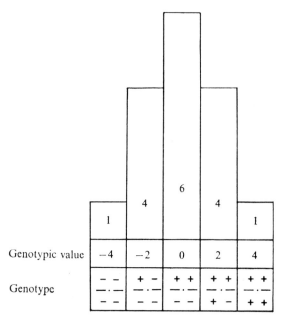

FIG. 5.2. Frequencies of genotypes from the cross $+/-.+/- \times +/-.+/-$ plotted according to the genotypic value (i.e., the relative number of $+$ and $-$ genes): the frequencies of each genotype are given in the histogram.

extremes, $+/+.+/+$ and $-/-.-/-$, with values of $+4$ and -4 respectively. If there is a third locus with two similar alleles, then for a cross between multiple heterozygotes the number of genotypic classes rises to seven, and with a fourth, to nine, and so on. The differences between the classes become progressively smaller as the number of segregating loci rises. At the stage when the differences between classes become about as small as the error of measurement, the distribution will become continuous, as in Fig. 5.1. In addition, any variation due to non-genetic causes will blur the underlying discontinuity implied by segregation, so that the variation seen may become continuous irrespective of the accuracy of measurement.

Thus an assumption of many genes, each with small effects, superimposed upon variability due to non-genetic or environmental causes, will lead to a continuous distribution similar to that given in Fig. 5.1. Genes which contribute to a quantitative trait, but which are not directly identifiable by classical Mendelian segregation (that is, cannot be studied individually) are referred to as *polygenes*, and genes whose effects can be studied individually are referred to as *major genes*. These categories do not imply any fundamental distinction between the two categories of genes. They are merely a matter of convenience, since the breeding methods used to study the effects of major genes cannot, in general, be used in the study of polygenes. Even so, it is possible under certain circumstances to magnify the effects of polygenes by statistical, and perhaps biochemical, techniques to such an extent that, to all intents and purposes, they behave as major genes.

The synthesis between the study of genes with observable qualitative effects and of those controlling continuously varying traits had its origins in a classic paper by Fisher (1918) and was further advanced by Wright (1921), Mather (1949), Kempthorne (1957), Falconer (1960), Mather and Jinks (1971), and others. A branch of genetics called *quantitative* or *biometrical* genetics has grown up to deal with the study of continuous variation.

Behavioural traits, such as duration of copulation in *Drosophila*, are essentially quantitative, and for their analysis an appreciation of the methods and aims of biometrical genetics is necessary. It should be emphasized again that the environment must be rigidly controlled for accurate results. A more detailed account of what is to follow appears in Parsons (1967a), and other general discussions are given on the applications of biometrical genetics to the study of the genetics of behaviour by Roberts (1967) and Broadhurst (1967).

The aim of biometrical genetics is to divide the phenotypic value which we measure into component parts attributable to different causes. The first division is into genotypic and environmental components. The genotype is the sum of the particular genes possessed by an individual, and the environment is the sum of all the non-genetic disturbances influencing the phenotypic value. We write for an individual: $P = G + E$, where P is the phenotypic value, G the genotypic value, and E the environmental deviation. The term 'environmental deviation' is used since we can think of the genotype as conferring a certain value on an individual, and the environment as causing a deviation from this in one direction or another. Arbitrarily the mean environmental deviation of the population as a whole is taken to be zero, so the mean phenotypic value (that is, the population mean) is equal to the mean genotypic value.

Since continuously varying traits are being considered, we need the phenotypic variance V_P, which is

$$V_P = V_G + V_E,$$

where V_G and V_E represent the genotypic and environmental variances. This relationship assumes that the genotype and environment are uncorrelated. If this is not so, the genotypic variance would be overestimated for a positive correlation between genotype and environment, and underestimated for a negative correlation. Correlations between genotype and environment can frequently be avoided in the laboratory, but may be important for behavioural traits, in which case we would need to estimate the degree of correlation. Assuming a correlation, the above equation becomes

$$V_P = V_G + V_E + 2\ Cov_{G,E},$$

where $Cov_{G,E}$ represents the covariance of genotypic values and environmental deviations, and provides, according to its magnitude, an estimate of the correlation between genotype and environment, or the genotype × environment interaction. It is one of the aims of biometrical genetics to develop methods for estimating V_G, V_E, and $Cov_{G,E}$.

Inbred strains maintained by brother–sister (sib) mating have been developed on many occasions in *D. melanogaster*. Sib-mating, being a form of inbreeding, leads to a progressive increase of homozygosity in each generation. In fact, the proportion of heterozygotes is expected to fall by 19·1 per cent per generation, so that in theory such strains will ultimately be completely homozygous. However, the approach to homozygosity may be retarded by the heterozygotes being fitter than the corresponding homozygotes. Assuming that complete homozygosity is attained, then all individuals *within* an inbred strain will be genetically identical, and all variation occurring will then be environmental. Between strains, there may be variation due to the different genetic constitutions of the strains as well as variation due to the environment. Even if several inbred strains are set up from the same population, the genetic constitution of the strains will be expected to differ, since by chance different loci are likely to be made homozygous in the different strains.

In Table 5.1 data on the duration of copulation in minutes are given for five inbred strains of *Drosophila melanogaster* which had been sib-mated in the laboratory for at least 140 generations (MacBean and Parsons 1966). In theory, they should be almost completely homozygous after this period

TABLE 5.1
Duration of copulation in D. melanogaster

Mean durations of copulation (minutes) based on 52 observations for each strain					
Strain	N1	N2	D5	G5	Y2
Mean (minutes)	19·27	16·69	21·42	19·50	17·65

Analysis of variance				
	d.f.	M.S.	F	E.M.S.
Between strains	4	172·75 M_1	12·34†	$V_E + 52V_G$
Within strains	255	14·00 M_2		V_E

† $P<0.001$, $V_E=14\cdot00$, $V_G=3\cdot05$, $h_B^2=0\cdot18$.

After MacBean and Parsons 1966

of time. The experimental procedure consisted of separating flies at eclosion, ageing them separately until they were three days of age, and then setting up single-pair matings. The time until mating commences is the mating speed, and immediately mating commenced, recording of the duration of copulation began.

An analysis of variance can be carried out to ascertain the relative importance of variation within and between strains (Table 5.1). The mean square (variance) within strains is represented by the error component in an analysis of variance. The mean square (variance) between strains is much greater than that within strains, and in fact the variance ratio F of $\dfrac{\text{M.S. between strains}}{\text{M.S. within strains}} = \dfrac{M_1}{M_2} = 12\cdot34$ ($P<0.001$). Thus, the variation between strains is significantly greater than that within strains. The within-strains M.S. represents the variation remaining after taking into account the strains, and is therefore the environmental variance V_E. Thus

$$V_E = M_2.$$

Between strains there are both genotypic and environmental variance components, and the expected mean square (E.M.S.) between strains can be shown to be

$$V_E + rV_G = M_1,$$

where r is the number of replicates for each strain, in this example being 52.

Thus, by equating the observed mean squares M_1 and M_2 with the expected mean squares we obtain

$$V_G = \frac{M_1 - M_2}{r} \text{ and } V_E = M_2.$$

The data give

$$V_G = 3 \cdot 05 \text{ and } V_E = 14 \cdot 00.$$

The total variance in the population, or phenotypic variance V_P, is equal to $V_G + V_E$, assuming no interaction between genotype and environment. It is then reasonable to compute the proportion of the phenotypic variance that is genotypic, thus

$$\frac{V_G}{V_G + V_E},$$

which is referred to as the *'heritability in the broad sense'* h_B^2. In our example $h_B^2 = 0 \cdot 18$. Clearly $0 < h_B^2 < 1$, for if $h_B^2 = 0$, then $V_G = 0$, and the trait would be determined entirely by the environment, and if $h_B^2 = 1$, then $V_E = 0$, so the trait would be determined entirely by the genotype. The figure of $0 \cdot 18$ is not particularly high but it does indicate a degree of genotypic control of duration of copulation.

It must be stressed that a heritability so estimated is a characteristic of the actual inbred strains under the environmental conditions prevailing. If the experiment were run under different conditions, or with different strains, or both, different values might be obtained. If we wish to make inferences about the components of variance and the heritability of a trait in a given population of an outbred species, we must set up inbred strains at random from the population. Then, in theory, the estimates obtained will relate to the parameters of the parent population rather than to the strains in the sample. This condition may not often be directly fulfilled but, to all intents and purposes, any set of inbred strains will provide useful estimates. However, in all cases the environment must be accurately defined. The procedure described may be extended to the study of a behavioural trait in a set of inbred strains over a series of environments (see Parsons 1967a) such as variations in temperature and light regimes. Such an analysis provides a comprehensive assessment of various genotype × environment interactions, some of which may be particularly important in behavioural work. It would permit a quick survey of the effects of a number of environments on a number of genotypes—an area of investigation which so far has been rather neglected in behavioural work.

2. The components of the genotypic variance and their estimation

The above experimental design takes no account of the genotype in terms of additive genes, dominance, and epistasis. Consider a single locus with two alleles A_1 and A_2. There are three genotypes at the locus A_1A_1, A_1A_2, and A_2A_2. Let us assign genotypic values $-a$, d, and $+a$ to these three genotypes respectively. We can then represent the genotypes on a linear scale (Fig. 5.3).

Genotype	A_1A_1		A_1A_2	A_2A_2
Genotypic value	$-a$	0	d	$+a$

FIG. 5.3. Genotypic values on a linear scale.

The zero point is equidistant between the two homozygotes. The value of the heterozygote, d, is positive if A_2 is dominant to A_1, negative if A_1 is dominant to A_2, and for no dominance, $d=0$. If $d > +a$, an *overdominant* situation is indicated. We wish to assess the relative contribution of genes with purely additive effects and those with dominance.

There may also be epistasis, or gene interactions, between loci. Taking these components of the genotypic value into account, and assuming no genotype × environment interactions, the genotypic variance can be written

$$V_G = V_A + V_D + V_I,$$

where V_A = variance due to additive genes (additive genetic variance), V_D = variance due to dominance deviations (dominance variance), and V_I = variance due to epistatic interactions (epistatic variance), so that

$$V_P = V_A + V_D + V_I + V_E.$$

Some experimental methods enable estimates to be made of all these components. Partly for simplicity, V_I is frequently ignored. Furthermore, where estimated, V_A and V_D are usually larger; however, most work is based on morphological traits, and certain behavioural traits could possibly present a different picture.

The ratio of the additive genetic variance to the total variance V_A/V_P can be computed, and is the '*heritability in the narrow sense*', h_N^2. It is a measure of the variation due to additive genes, and is more useful than the heritability in the broad sense defined previously as $\dfrac{V_G}{V_P}$, because when considering relationships between generations, it is the gametes carrying genes rather than genotypes that are passed on from one generation to the next. Thus from the predictive point of view, the ratio $\dfrac{V_A}{V_P}$ is more useful. A number of experimental designs are available for estimating V_A and V_D. The literature on models and experimental designs is extensive, therefore it is proposed to look briefly and only descriptively at a few of the available techniques likely to be of use in behaviour-genetic analysis.

In one method, two homozygous inbred strains which can be referred to as parental strains P_1 and P_2 are taken, and various crosses are carried out to give the F_1, F_2, B_1 the back-cross of the F_1 to the parent P_1 ($P_1 \times F_1$), and B_2 the back-cross of the F_1 to parent P_2 ($P_2 \times F_1$). A total of fourteen generations can, in fact, be set up if reciprocal crosses between parents are taken, and all other possible crosses are carried out reciprocally. From such an analysis, the genotypic variation can be separated into additive effects, dominance, epistasis, and reciprocal differences where sex-linkage can be distinguished from maternal effects—a form of environmental influence common to the behaviour of many species. Details of this technique are presented with its theoretical justification by Mather (1949) and in subsequent papers by Mather and co-workers (see Mather and Jinks 1971).

The method permits various scaling criteria to be set up before the detailed partitioning of the genotypic variation into its components. The question of scaling is, in fact, considered by many to be one of the most intractable problems of biometrical genetics. Different conclusions may well emerge when different scales are used, and interactions can be radically altered by transforming the data.

This method has been rather little applied to behavioural genetics. However, Kessler (1969), using lines selected for fast and slow mating speeds in *D. pseudoobscura* at the fifteenth generation of selection, obtained $h_N^2 = 0.25$ for the performance of groups of flies, and, when they were tested as single pairs, $h_N^2 = 0.17$. In both cases, dominance of genes in the fast-mating lines was found.

As pointed out by Parsons (1967*a*), if we can assume the parentals P_1 and P_2 to be completely homozygous, it is possible to test all generations at the same time by using the parents in two successive generations and the F_1 in one. When the likely importance of environment is under consideration, this may be of some significance as complications due to environmental variations between times would be avoided. Furthermore, specific environments can be studied, especially given the likely importance of genotype × environment interactions for behavioural traits. The need to carry out these experimental designs in a multiplicity of environments, especially temperatures, will become apparent in later chapters.

A very powerful technique consists of the procedure just discussed used in association with the diallel cross, and we therefore turn to a consideration of diallel crosses.

3. Diallel crosses

Probably the most useful technique so far in behaviour genetic work is the diallel cross. This is the set of all possible matings between several strains or genotypes; thus, for four strains there are sixteen possible combinations:

Strain of male parent

		A	B	C	D
	A	AA	AB	AC	AD
Strain of female	B	BA	BB	BC	BD
parent	C	CA	CB	CC	CD
	D	DA	DB	DC	DD

made up of six crosses AB, AC, AD, BC, BD, and CD, their six reciprocals BA, CA, DA, CB, DB, and DC, where the sex of the parents is transposed, and four kinds of offspring derived from the four parental strains AA, BB, CC, and DD which are arranged along the leading diagonal. In general, if there are n strains, the diallel table will have n^2 entries made up of $\frac{n(n-1)}{2}$ crosses, $\frac{n(n-1)}{2}$ reciprocals, and n parental strains. Diallel crossing techniques vary depending on whether the parental strains or the reciprocal F_1's are included, or both.

There is a variety of theoretical methods of analysis of diallel crosses, depending somewhat on the information required (see, for example, Mather 1949, Hayman 1954, Jinks 1954, Griffing 1956, Kempthorne 1957, Wearden 1964, Mather and Jinks 1971). The first detailed genetic analysis of a diallel cross for a behavioural trait (Broadhurst 1960) was carried out on six strains of rats for defecation scores (number of faecal boluses deposited in an arena in exactly two minutes) and ambulation scores (number of marked areas entered in the arena in exactly two minutes), using the analytical methods of Mather (1949) and his colleagues. Griffing (1956) presented a method which, by a relatively simple analysis of variance, yields information on the additive, dominance, and environmental components of variance of a trait. As for the previous methods of analysis, the validity of the estimates of the variance components depends on the strains being assumed to be a random sample from some population about which inferences are to be made. However, in some cases we cannot assume this as the strains may be deliberately chosen, and so the experimental material then constitutes the entire population about which inferences are to be made.

Parsons (1964) carried out a complete 6×6 diallel cross for mating speed in *D. melanogaster* made up of six inbred strains that had been sib-mated for at least 120 generations at the time of the experiment, and the thirty possible hybrids between them. Males and females were stored separately at eclosion, and at six days of age a single male was shaken with a female and observed until copulation began. The time in minutes for this to occur is the mating speed. Pairs not mating in $\leqslant 40$ minutes were recorded as unmated. For

TABLE 5.2

Mean numbers of successful matings out of 7 in 40 minutes in D. melanogaster

		\multicolumn{6}{c}{Strain of ♂ parent}

		N1	N2	Y1	Y2	G5	OR
Strain of ♀ parent	N1	*4·5*	5	6	0·5	6	7
	N2	6	*6·5*	5·5	5	6	7
	Y1	6·5	7	*2·5*	6	4·5	6·5
	Y2	2·5	6·5	4·5	*0·5*	7	7
	G5	5·5	5	5	3	*6·5*	7
	OR	7	6	6·5	6	6	*3·5*

After Parsons 1964

each of the thirty-six possible combinations, fourteen trials were carried out split into two replicates of seven. In Table 5.2 the mean numbers mating out of seven trials are given for those mating in 40 minutes. The means for the inbred strains are indicated in italic. In Table 5.3 the means of the females of a given strain with the five other strains $X_{i.}$, and the means of the males of a given strain with the five other strains $X_{.i}$, are given. The means for the inbred strains themselves X_{ii} are also given. Results are presented for three time intervals, namely, those involving mating in $\leqslant 10$, $\leqslant 20$, and $\leqslant 40$

TABLE 5.3

Mean numbers of successful matings for hybrids $X_{i.}$ and $X_{.i}$ and inbred strains X_{ii} in D. melanogaster

Strain	\multicolumn{9}{c}{Time period} ♂

	$\leqslant$ 10 minutes			$\leqslant$ 20 minutes			$\leqslant$ 40 minutes		
	$X_{i.}$	$X_{.i}$	X_{ii}	$X_{i.}$	$X_{.i}$	X_{ii}	$X_{i.}$	$X_{.i}$	X_{ii}
N1	2·4	3·5	1·5	3·1	4·7	2	4·9	5·5	4·5
N2	2·9	3·5	4·5	4·4	4·6	5	5·9	5·9	6·5
Y1	4·1	2·6	2	5	3·7	2	6·1	5·5	2·5
Y2	3·9	2·4	0·5	4·9	3·2	0·5	5·5	4·1	0·5
G5	2·7	3·6	1·5	3·9	4·5	3·5	5·1	5·9	6·5
OR	5·2	5·6	1	6	6·6	2	6·3	6·9	3·5
Over-all mean	3·53	3·53	1·83	4·55	4·55	2·5	5·63	5·63	4

After Parsons 1964

minutes. Significant hybrid vigour is revealed by comparison of the inbreds and hybrids, especially for the $\leqslant 10$-minutes data. That the differences are lower at the 40-minutes time interval is reasonable, since the maximum values possible are being approached. Although other examples of hybrid vigour will be mentioned, it is of interest that Bösiger (1960) has shown that the degree of inbreeding and reduction of sexual vigour are positively correlated in *D. melanogaster*, and Maynard Smith (1956) demonstrated that inbreeding in *D. subobscura* resulted in decreased athletic ability. The inbred male was not able to follow the female successfully as she moved from side to side, during one part of the male courtship pattern.

Before detailed analyses were carried out on the data in Table 5.2, the angular transformation was applied, as this removes a dependence of the variances on the means. The inbred strains data were first analysed using the type of analysis of variance given for the duration of copulation data in §5.1. Differences between strains were significant at the 0·1 per cent level for the $\leqslant 10$-minutes data, and at the 1 per cent level for the $\leqslant 20$- and $\leqslant 40$-minutes data. Heritabilities in the broad sense h_B^2 came to 0·81, 0·73, and 0·75 for the three time periods respectively. Thus, there are substantial differences in mating speed between the inbred strains. Estimates of V_A and V_D cannot be obtained from the inbred strains themselves, but can be obtained from the hybrids. The further analysis of the diallel cross will be based on the hybrids without the inbred strains, because Griffing (1956) has recommended that to obtain unbiased estimates of V_A and V_D the inbred strains should be omitted.

From Table 5.3, an over-all idea of a given strain in hybrid combination with all the other strains can be obtained from $X_i.+X._i$. For the $\leqslant 10$-minutes data the extreme values are 5·9 for strain N1 and 10·8 for OR. The average mating speed of a given strain in hybrid combination, obtained by considering all possible hybrids of that strain, is referred to as the *general combining ability* (g.c.a.) of that strain. Normally, general combining abilities are expressed as deviations from the over-all population mean. In any case, it is clear that the g.c.a. of N1 < OR. Given the general combining abilities of two strains, say N1 and OR, it is possible to predict an expected average mating speed of the crosses between N1 and OR, that is for N1♀×OR♂ and its reciprocal. The degree to which the mating speed of this cross differs from that predicted on the basis of the general combining abilities of N1 and OR is referred to as the *specific combining ability* (s.c.a.). Finally, it is possible to calculate the degree to which N1♀×OR♂ differs from its reciprocal OR♀×N1♂, and this is referred to as a *reciprocal effect*.

Table 5.4 gives the results of analyses of variance of the proportions mating in $\leqslant 10$, $\leqslant 20$, and $\leqslant 40$ minutes, showing highly significant g.c.a.'s for all time periods, while the s.c.a.'s and reciprocal effects are significant but at a lower level (see F values for model I). As might be expected from Table 5.3,

TABLE 5.4

Analyses of variance of the diallel cross of hybrids for the mean number of successful matings in D. melanogaster

		Model I (for differences between strains)		
			Time period	
		>10 minutes	>20 minutes	>40 minutes
	d.f.		F values	
G.c.a.	5	13·42†	16·94†	9·51†
S.c.a.	9	3·04‡	4·82†	4·15§
Reciprocal	15	3·02§	3·56§	2·63‡
Error	30			
		Model II (for components of variance)		
G.c.a.	5	5·07‡	3·51‡	2·29
		Variance components		
	V_A	256	228	106
	V_D	85	144	124
	V_E	166	150	158
	h_N^2	0·51	0·44	0·27

† P<0·001, ‡ P<0·05, § P<0·01.
The details of differences between the two models are given by Griffing (1956); in model II the F values for s.c.a. and reciprocal effects are as for model I.

After Parsons 1967a

N1 has a significantly positive g.c.a., and Y2 and OR significantly negative g.c.a.'s. Various s.c.a.'s were significant, the most significant being a negative s.c.a. for the hybrid between N1 and Y2. Since we are dealing with a series of inbred strains and their hybrids, it is possible to obtain estimates of V_A, V_D, and V_E from this analysis and hence h_N^2 (model II in Table 5.4). Values of h_N^2 came to 0·51, 0·44, and 0·27 for the ⩽10-, ⩽20-, and ⩽40-minutes data respectively. Again, the lower heritability after the longer time interval occurs presumably because we are closer to the maximum possible number of matings than after 10 minutes. This technique, therefore, although showing a high level of genotypic control over short time intervals, makes interpretations difficult because of the dependence of the conclusions on the actual time intervals. The other difficulty is that mating speed measured in both sexes at once relates to the behavioural interaction of a particular pair of genotypes, and so generalization to other possible genotypic combinations, such as would certainly occur in natural populations, is difficult.

Fulker (1966) used a technique which, by focusing attention on the males alone in *D. melanogaster*, provides a reliable estimate of heterosis, variance components, and heritability. Single males from each of six inbred strains were tested with six virgin females (one female from each inbred strain) over a period of 12 hours. Since each male was given the same array of females to mate with, the females can be considered as a standard testing set, and in comparison with Parsons' (1964) data, it is clear that we are dealing with the effect of one sex alone. Table 5.5 shows that the time to the first copulation and the observed number of copulations were much more uniform for hybrids than for the inbreds. Correlations at the foot of the table indicate that the males which mate quickly on the first occasion also copulate more often, more successfully, and leave more progeny. The four measures thus appear

TABLE 5.5

Mean scores (based on 5 males each tested with 6 females) for male genotypes

Genotype	Time to first copulation† (A)	Observed number of copulations (B)	Number of copulations resulting in fertilization (C)	Number of offspring produced (D)
Inbred strains				
Edinburgh (Ed)	9·2	4·6	4·0	147
6CL	34·6	1·6	1·0	33
Samarkand (S)	26·4	2·8	1·8	98
Wellington (W)	16·4	4·2	3·6	236
Oregon (Or)	18·2	2·8	2·4	158
Florida (F)	14·8	3·6	2·8	118
Hybrids				
F × Ed	3·6	6·2	5·4	302
F × W	6·8	5·8	5·4	340
Or × W	5·6	5·2	5·0	270
6CL × Ed	5·6	5·0	4·6	234
6CL × S	7·6	3·2	1·4	63
Or × Ed	5·2	5·4	4·2	225
Intercorrelations	AB −0·87 ($P<0·001$)			
	AC −0·78 ($P<0·01$)			
	AD −0·69 ($P<0·02$)			
	BC 0·96 ($P<0·001$)			
	BD 0·90 ($P<0·001$)			
	CD 0·95 ($P<0·001$)			

† Those not mating in 40 minutes were assigned a value of 41.

After Fulker 1966

as aspects of a general characteristic of male mating behaviour. The intercorrelations further suggest that the number of copulations resulting in fertilization is a convenient method of measuring mating speed, since with $r=0.96$ for the correlation between this and the observed number of copulations, the two measures are practically interchangeable. Thus a 6×6 diallel cross was set up based on the number of copulations resulting in fertilization in 12 hours, which had the advantage of avoiding the need for constant observation over a 12-hour period. The diallel cross was analysed by the rigorous analytical methods of Hayman (1954) and Jinks (1954) which yield five parameters; one measuring additive genetic variation, two being dominance variation parameters differing only in the presence of unequal gene frequencies, another indicating whether dominant or recessive alleles are more frequent, and a final parameter representing environmental variation. Conveniently, it turned out that the only parameters that were relevant were V_A, V_D, and V_E, and no reciprocal effects were found. A heritability in the narrow sense of 0.36 was obtained from the diallel cross. A high level of heterosis for number of females fertilized was also shown. The diallel showed strong directional dominance for high frequency of mating—a conclusion coming from a graphical technique described by Hayman (1954), Jinks (1954), and others, but which is too complex to give here (see, however, Mather and Jinks (1971) for a description of these approaches). Fulker argues that this result implies a history of natural selection for the maximum value of the trait rather than an intermediate or low value—that is, the trait is subject to directional selection. Independent evidence for the importance of high mating speed as a component of fitness is provided by the high correlation with yield ($r=0.90$) found in the first experiment. In agreement are data of Kessler (1969) in *D. pseudoobscura* which showed directional dominance for fast mating, although other fitness factors were not analysed. Prakash (1967) also showed mating speed to be a component of fitness in *D. robusta*, since he found an association of mating speed with repeat matings, and with fertility in males. For females, repeat mating is less likely since in *Drosophila* a certain refractory period always follows a mating; and in any case repeat mating in females does not increase their productivity if the second mating comes before any sperm have been utilized from the first mating.

Hay (1972a) studied two behavioural traits of a non-sexual nature, namely, activity and preening as responses to mechanical stimulation. Using a method which allowed large numbers of individuals to be tested rapidly, diallel crosses of six inbred strains, together with the fourteen generations, as described in §5.2, derived from the extreme parental lines Edinburgh and 6C/L, were carried out. This type of analysis, where the diallel cross is supplemented by the fourteen generations, has many advantages as an intensive method of biometric analysis (Cavalli 1952). It enables the separation of genetic variation into additive effects, dominance effects, epistasis representing additive and dominance components, and reciprocal effects where sex linkage can be distinguished from maternal effects. The analyses carried out by Hay are the most comprehensive carried out so far from the biometrical point of view, and demonstrate clearly the value of such techniques. Further, since they concern non-sexual traits the problem of interactions between sexes does not occur.

Hay (1972a) found that the genetical control of activity and preening cannot be regarded as simple, as it varied between different experimental conditions. Thus, if adequate food were available in the seven-day period between eclosion and testing, dominance for high activity and low preening was found. The dominance for high activity was enhanced by duplicate interaction, suggesting that there had been directional selection for the trait in the direction of high activity. This was confirmed in a less favourable environment, including a restriction of the food supply, which led to an increase in mortality chiefly in the least active strains, although there was an over-all rise in activity due to a higher death rate among the less active flies in all lines. For preening, dominance was found only when the density was low, suggesting directional selection for low preening. In high-density cultures, however, where preening must serve more and more to fend off other flies (see Sexton and Stalker 1961) so that they are distributed over the available food and egg-laying space, the only significant components were the additive and the dominance maternal effects, which buffered any autosomal contributions and had the effect of reducing the range of variation between the strains.

Hay (1972b) found genotype × environmental interactions over the first 15 days of adult life. Activity declined over the 15 days and a linear interaction of the additive genetical component led to heterosis for high activity at day 15 as compared with day 1 (Table 5.6). Preening, on the other hand, did not vary with age, but like activity, showed replicate interactions traceable to one inbred strain and smaller inter-individual variances in the F_1's than in the parents. As already pointed out (§4.2), heterozygotes tend to show

TABLE 5.6

Activity in tubes on days 1 and 15 expressed as the proportion in angles of all responses (each statistic is based on 120 flies)

Genotype	Day 1	Day 15
Edinburgh (E)	$51·88 \pm 1·41$	$29·99 \pm 1·48$
6C/L (C)	$64·82 \pm 0·96$	$29·65 \pm 3·70$
E × C	$61·54 \pm 1·15$	$44·18 \pm 1·59$
C × E	$63·84 \pm 1·08$	$38·83 \pm 1·55$

After Hay 1972b

less variability over environments for some quantitative traits than do homozygotes. Such issues are discussed further in Chapter 13. It is clear, however, that the techniques of biometrical genetics are of considerable use in the assessment of genotype × environment interactions for behavioural traits. In fact, because of the dependence of many behavioural traits on environmental variables, the rather sophisticated techniques now available in biometrical genetics should prove powerful in behaviour-genetic analysis.

Hay has therefore combined the techniques of diallel cross analysis with those described in §5.2, in an effort to look further into the genetic architecture of behavioural traits than has been so far attempted. As is shown in Chapter 6, the alternative approach to the analysis of quantitative traits, which is based on the directional selection experiment, offers the possibility of actually locating and studying the genes controlling a trait; it would be of interest to use this approach on Hay's behavioural traits, since there is clearly a limit to the amount of information the biometrical approach can yield, partly because of some of the basic assumptions made in the setting-up of models.

Finally, the diallel cross provides a useful method of ascertaining the sex important in controlling mating speed, as shown in Chapter 4 for 3×3 diallel crosses between karyotypes of *D. pseudoobscura*. In *D. melanogaster* Parsons (1965*a*) considered a 5×5 diallel cross of inbred strains themselves (not their hybrids). The number mating out of eight in 10 and 40 minutes was found to be under genetic control. At the 10-minutes period the strain of the male was highly significant ($P<0.001$), while at the 40-minutes period less so ($P<0.05$). However, the strain of the female approached significance at this stage. Although different total numbers of matings were involved for the two time periods, it seems that initially the vigour of the males leads to some rapid matings, whereas at a later stage variations in female receptivity may become relatively more important.

Therefore, in *D. melanogaster* from the diallel scheme the initial matings seem male-dependent, but as the observation time is increased the female's receptivity seems to become progressively more important, as discussed earlier. In *D. pseudoobscura*, mating is so rapid that variations in the female's receptivity may be relatively unimportant. In *D. persimilis*, a sibling species of *D. pseudoobscura*, it has been shown that females are critical over a one-hour period because of an interaction between copulation and avoidance tendencies (Spiess and Langer 1964*b*). Thus the sex more important in determining mating speed varies between species, and within species between genotypes, such that (at least in *D. melanogaster*) the male may often be more important if mating is rapid, while if mating is slower the female plays a progressively more important role. Unfortunately, it is still difficult to generalize, since few karyotypes and genotypes have been studied in detail, and differences in experimental details are probably very important in these experiments. Certainly, in *D. melanogaster*, experiments studying the mating behaviour of flies without antennae and wings show the importance of the antennae of females and the wings of males in courtship behaviour (Mayr 1950, Bastock 1956, and see §3.1), and it is likely that the odour of both sexes is relevant (see §7.2), that of females being more effective than that of males in *D. melanogaster*.

Hay (1972c) has modified the method of analysing the diallel cross to study behavioural interactions between individuals or between sub-groups, even if it is impossible to identify the participants in an encounter. This was applied to six inbred strains of *D. melanogaster*, tested in all possible pairwise combinations. The presence of two different strains in the same group enhanced the normal strain differences in preening behaviour, while the spacing of the flies became less random in that the mean distance between individuals was similar in all groups consisting of two strains. For both measures the interaction between strains was most pronounced among flies left in the original culture bottle for at least ten days before testing. Such an environmental effect suggests that a specific colony odour may be involved in the recognition by *Drosophila* of other genotypes, and is supported by evidence for discrimination between flies from different cultures of the same strain. He proposes that such olfactory mechanisms may be of importance in phenomena such as frequency-dependent mating (see §7.2).

In conclusion, from the methodological point of view, the diallel cross can be used for estimating the variance components of behavioural traits, and for studying behaviour itself for combinations of various strains. It is a powerful technique for a general survey of a series of strains, perhaps in several different environments, whatever the aim or model used. It has the advantage that a single generation provides a great deal of information, although some additional information can come from further generations. For behavioural traits where genotype×environmental interactions may be important, the one-generation approach has clear advantages. The diallel technique is an extensive analytical method rather than intensive, as a number of strains and hybrids can be surveyed at once. It represents an eminently suitable first technique in an investigation of a behavioural trait in an organism such as *D. melanogaster* where adequate inbred strains exist, and, as discussed earlier, it can provide much information from which more specific experiments may be carried out, such as those described in the next chapter.

4. Relationship of offspring to parent

For many organisms no ready supply of inbred strains is available. This would apply to the species of *Drosophila* not yet studied in detail in the laboratory. By working out correlations between relatives estimates of heritabilities may be obtained: in the treatment of resemblances between relatives, we assume the parents to be a random sample of their generation and to be mated at random. Comprehensive accounts are given by Falconer (1960) and in other texts. Here we will merely consider relationships between parent and offspring. It is not difficult to visualize that half the genes of any offspring will be common with a given parent and half will be different. Thus, of the additive variance, V_A, in the parents, half will go to the offspring, so that the covariance between parent and offspring $=\frac{1}{2}V_A$. Now since the phenotypic variance, V_P, in the parental and offspring generations can be assumed to be the same, the correlation coefficient between parent and offspring

$$r_{OP} = \frac{\frac{1}{2}V_A}{\sqrt{V_P \cdot V_P}} = \frac{1}{2}h_N^2.$$

Given data on both parents, the correlation between mid-parent and offspring can be computed. Writing the mid-parental variance as $V_{\bar{P}}$, where $\bar{P} = \frac{1}{2}(P_1 + P_2)$, assuming the phenotypic variance to be the same in both parents and P_1 and P_2 are the values of the two parents, gives $V_{\bar{P}} = \frac{1}{2}V_P$, so that the correlation between mid-parent and offspring becomes

$$\frac{\frac{1}{2}V_A}{\sqrt{\frac{V_P \cdot V_P}{2}}} = \frac{h_N^2}{\sqrt{2}}.$$

For relationships between parent and offspring a regression approach can also be used. It is well known from the work of Galton and Pearson that the sons of tall men tend to be tall, but not as tall as their fathers, yet not as short as the average of the population: in fact, the height of sons tends to be about half-way towards that average. Similarly, the sons of short men tend to be short, but not as short as their fathers, and on average have heights about half-way between their fathers and the population average. This regression by one-half towards the mean is exactly what we would expect based on additive genes. In fact, if b_{OP} is the regression coefficient of one offspring on parent, it can be shown that

$$b_{OP} = \frac{1}{2}\frac{V_A}{V_P} = \frac{1}{2}h_N^2 = r_{OP},$$

where r_{OP} is the correlation between parent and offspring. The better parent to use is the father, since the regression of offspring on the mother may give too high an estimate because of maternal effects.

It can also be shown that the regression of offspring on mid-parent, $b_{O\bar{P}}$, is equal to the heritability, h_N^2, or

$$b_{O\bar{P}} = h_N^2.$$

The use of relationships between mid-parent and offspring depends on the variances being equal in both sexes; hence, it is perhaps of less service than the relationships between one parent and offspring. Furthermore, so far as behaviour is concerned, many traits are sex-limited—such as components of sexual behaviour—so that the mid-parent approach would be invalid. Sometimes the mating of parents is not at random but according to their phenotypic resemblance, a system known as *assortative mating*. There is then a correlation between the phenotypic values of the mated pairs. The consequences of assortative mating have not been studied in great detail, and in practice are often ignored or perhaps circumvented by the experimental design, but in situations where mating cannot be controlled its existence must be remembered. Its importance will depend, among other things, on the heritability of the trait under study. It is merely mentioned here as a possible source of bias (see also Chapter 7).

Connolly (1966) studied locomotor activity in *D. melanogaster* by measuring activity in an open-field apparatus, as assessed by the number of squares an insect crosses each minute. An estimate of heritability was obtained from the regression of offspring on mid-parent. Twenty-five pairs of parents selected from a wild-type stock (Pacific strain) were mated in single pairs. From among the offspring of each of these matings, two were selected at random and measured. Fig. 5.4 shows the regression of offspring on mid-

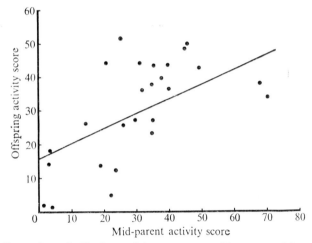

FIG. 5.4. Regression of offspring activity scores on mid-parent activity scores. (After Connolly 1966.)

parent. The equation of the linear regression is $Y = 15.56 \pm 0.51\ X$. The slope of the line estimates the heritability, since we are dealing with mid-parental values. Therefore $b_{OP} = h_N^2 = 0.51 \pm 0.10$, which is significantly >0 ($P<0.01$). For out-bred strains, therefore, this procedure provides a fairly rapid estimate of heritability. Further discussions of Connolly's results are given in §6.4, where experiments on selection for locomotor activity are described.

5. Conclusion

The methods described are not unique, being merely routine methods of biometrical genetics as adapted to behaviour genetics. Many of the techniques, except for those given in §5.4 have been used more extensively in plant than in animal breeding. At the behavioural level, animals tend to be much more environment-sensitive than at the morphological level, hence the usefulness of techniques which are aimed at detecting, estimating, and controlling environmental effects as developed in plant breeding (Caspari 1968, Parsons 1967a). Much of the work reported so far has been on mating behaviour, but in the next chapter on selection experiments a greater diversity of traits is considered. It is perhaps unfortunate that there has been a concentration on mating behaviour for the application of the techniques discussed in this chapter, because of the possibility of complications in the interpretation of variance components and heritabilities caused by interactions of a behavioural type between mating individuals. For mating behaviour, this is best avoided in the type of design used by Fulker (1966), where he specifically studied males against a constant array of females. Future work on other behavioural traits in *Drosophila* using biometrical techniques will be awaited with interest, and the start made by Hay (1972a, b) is encouraging.

6

SELECTION EXPERIMENTS

1. Introduction

THE selection experiment arises out of the biometrical approach just discussed, and consists of manipulating the genotype by selecting and mating various chosen phenotypes with respect to a trait, from a population. We shall be mainly concerned with the form of selection called *directional* selection, where extreme individuals of a population are selected at the expense of all others in the hope of forming separate high or low lines in subsequent generations. Two other forms of selection are mentioned in various contexts in this book, namely, *disruptive* selection, in which selection is practised at both the high and low extremes of the distribution in each generation, and *stabilizing* selection, in which intermediate phenotypes are selected at the expense of both extremes (Fig. 6.1).

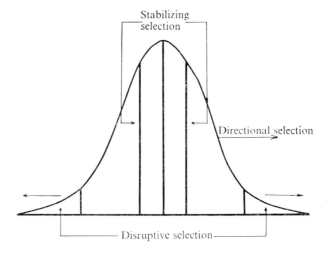

FIG. 6.1. The sections of a normal distribution favoured under directional, disruptive, and stabilizing selection.

In directional selection, if a quantitative trait has some genetic basis there should be a response to selection, since the selection of extreme phenotypes means that extreme genotypes will also be selected. At the outset at least, the magnitude of the response R to selection will depend on two parameters;

the heritability in the narrow sense, h_N^2, and the selection differential, S, which is the mean phenotypic value of the individuals selected as parents expressed as a deviation from the mean phenotypic value of all individuals in the parental generation before selection was begun. The magnitude of S depends on the proportion of the population included among the selected group and the standard deviation of the trait (see Falconer 1960). Provided that fertility and viability are not correlated with the trait, the response to selection is given by

$$R = h_N^2 S.$$

Clearly if $h_N^2 = 0$, no response is possible since the trait would be determined entirely environmentally, and the larger the heritability the greater the response. In theory the prediction of response is valid for only one generation, because the basic effect of selection is to change gene frequencies and hence the genetic properties of the offspring. However, in many experiments the predicted response is maintained for five or more generations. A great deal of effort and statistical sophistication has been devoted to the question of the estimation of the predicted response, which, of course, depends on the accurate estimation of h_N^2 by some of the methods already discussed. If the heritability had not been estimated before selection, the above equation provides an estimate of it as R/S.

During the process of selection, extreme phenotypes are continuously favoured. This, in all likelihood, will lead to an increasing proportion of extreme and probably homozygous genotypes. This will be accelerated if the sizes of the populations used as parents are extremely small because inbreeding, and hence homozygosity, will occur, reducing the level of variability on which selection can act. Thus, except for extremely large population sizes, this limits the number of parents and so limits S. Ultimately, probably because of increasing numbers of homozygotes, the rate of response to selection will be expected to diminish, and perhaps there will be a series of generations when there is no response—that is, the population will be at a plateau. The rate of response may also diminish if general fertility and viability decrease during selection. Even so, occasionally after some generations at a plateau there may be a rapid response for a few generations, which is called 'an accelerated response to selection'. To a certain extent these rapid responses have been shown to be repeatable (Thoday and Boam 1961) in selection experiments for sternopleural chaeta number in *D. melanogaster*. This argues against contamination as a cause, and population sizes are usually far too small in selection experiments to invoke mutation. The likely interpretation is recombination between linked genes controlling the trait leading to extreme gametes, which then, being favoured by selection, will increase in frequency rapidly, probably leading to extreme homozygotes, unless the homozygotes are highly unfit or lethal, as may occasionally happen.

FIG. 6.2. Photograph of a 15-unit maze in vertical position. Flies are introduced in the vial at the left and are collected from various vials at the right.

If selection leads to an increased proportion of homozygotes, or at least to a fall in genetic heterogeneity, then there is the possibility of assessing the contribution of each chromosome to the response to selection in organisms such as *D. melanogaster*, where suitable stocks exist to aid in such an analysis. It may be possible to go so far as to identify and locate actual genes responsible, as has been done by Thoday (1961) and his colleagues in analysing the response to selection for sternopleural chaeta number.

In this chapter some of the *Drosophila* experiments which show responses to selection for behavioural traits will be discussed. There is the possibility of learning something not only of the genetic basis of behaviour, but also of behaviour itself, especially if the behavioural trait under analysis can be divided into components, some of which may be differentially affected by selection. (For more details of the biological basis of selection, the review of Lee and Parsons (1968) may be consulted.)

2. Examples (mainly genetic analysis)

Perhaps the most complete analysis from the genetic point of view is that by Hirsch and Erlenmeyer-Kimling (1962) of geotaxis (gravity-oriented locomotion) in *D. melanogaster* (see also Erlenmeyer-Kimling, Hirsch, and Weiss 1962, Hirsch 1963, 1967). A vertical 10- or 15-unit plastic maze was used (Fig. 6.2), into which flies are introduced on the left-hand side of the maze and are collected from one of the vials on the right. They are attracted through the maze by the odour of food, and by lighting in the form of a fluorescent tube on the right-hand side. The apparatus permits the screening of the behaviour of a large number of flies with the exclusion of human handling once the flies have been introduced; thus, conditions for maximum objectivity are provided. Selection for positive and negative geotaxis produced rapid and clear responses (Fig. 6.3).

Erlenmeyer-Kimling assayed the role of the three major chromosomes of *D. melanogaster* in the response to selection for geotaxis using the methods described by Mather (1942) and Mather and Harrison (1949) in chaeta number experiments. By the use of stocks carrying dominant genes and also inversions inhibiting crossing-over in the relevant chromosomes, it is possible to combine in all possible ways chromosomes extracted from a selection line with those from the control line, and so study the individual effects of the chromosomes and their interactions, with respect to a quantitative trait such as chaeta number or geotaxis.

A multiple tester stock is therefore taken, for example: $A/+$, B/C, D/E, where A, B, C, D, and E are dominant markers on the X, II, and III chromosomes respectively. The presence of inversions in the tester chromosomes prevents crossing-over in heterozygotes, so that they segregate to all intents and purposes as whole units. Females of this stock are crossed to males of

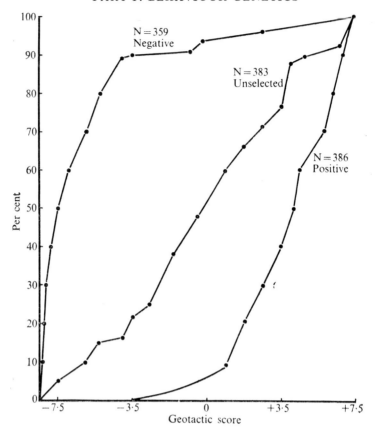

FIG. 6.3. Cumulative percentages of flies (males and females) for various geotactic scores in a 15-unit maze, for an unselected foundation population and two selected lines. (After Hirsch 1963.)

the line to be assayed, giving females of the type A/S, B/S, D/S, where S represents the selection line to be assayed. These females are then back-crossed to the selection line under analysis. Assuming markers A, B, and D to be associated with inversions, the S chromosomes in the above females will be broken down very little by recombination, and thus they will go to the next generation relatively intact. In the back-cross there are eight classes of progeny, namely,

(1)   A   B   D
(2)   A   B
(3)   A   D
(4)   A
(5)   B   D

(6) B
(7) D
(8) selected line.

These eight classes thus receive a representative of each S chromosome through the father, and from the mother a variable number of S chromosomes or tester chromosomes T (A, B, D). Thus, each of the major chromosomes are heterozygous or homozygous for an S chromosome (in females). The eight classes make up all possible combinations between T chromosomes. By using females, the individual effects of the chromosomes and their interactions can be studied for a quantitative trait, whether it be chaeta number or a behavioural trait such as geotaxis. Furthermore, since there is no crossing-over in the male in *Drosophila*, whole chromosomes may be maintained indefinitely by back-crossing males heterozygous for the S chromosomes, and for a recessive marker on each chromosome, to females homozygous for the same recessive marker. One limitation of the technique is that it is only fully efficient in detecting recessive genes in the S chromosomes, since the comparison for each chromosome is between the heterozygote T/S and the homozygote S/S. Genes on the selected chromosome S completely dominant to their corresponding allele on the T chromosome will not be detected, and genes partially dominant will be detected with a variable degree of efficiency, according to the degree of dominance (see Hirsch 1962).

TABLE 6.1

Mean chromosomal effects (with standard errors) on geotactic scores after selection based on a unit maze: in each case the means are based on 10 replicates

Population	Chromosome		
	X	II	III
Selected for positive geotaxis	1.39 ± 0.13	1.81 ± 0.14	0.12 ± 0.12
Unselected	1.03 ± 0.21	1.74 ± 0.12	-0.29 ± 0.17
Selected for negative geotaxis	0.47 ± 0.17	0.33 ± 0.20	-1.08 ± 0.16

After Hirsch 1967

Table 6.1 gives the mean effects of chromosomes X, II, and III in the unselected population, and in those selected for positive and negative geotaxis. In the unselected population, the X and II chromsomes produced positive, and chromosome III negative, geotaxis by comparison with the standard tester chromosome. Selection for positive geotaxis had little, if any, effect on chromosome II, but might have increased the positive effect on the X, and changed a negative effect on chromosome III to slightly positive. In general, selection for negative geotaxis had a greater effect on all three chromosomes than did selection for positive geotaxis. In any case, selection was far more effective for negative than for positive geotaxis (Erlenmeyer-Kimling et al. 1962). The analysis taken this far thus shows that there are genes distributed on the three major chromosomes of Drosophila affecting gravity-oriented locomotion.

Spassky and Dobzhansky (1967) found that geographic strains of D. pseudoobscura and D. persimilis showed considerable variability for geotactic response, the modal behaviour in both species being geotactic neutrality. It is therefore not unreasonable that in selection experiments on geotaxis in D. pseudoobscura with the same type of maze as described earlier (Dobzhansky and Spassky 1962), selection in both positive and negative directions was effective in populations monomorphic as well as polymorphic for the third chromosome arrangements AR and CH. In polymorphic populations, it was found that selection for positive geotaxis favoured the carriers of AR chromosomes, while that for negative geotaxis gave an advantage to the AR/CH heterokaryotypes. Relaxation of selection led to a partial relapse towards the original state, indicating that the gene pool had not been made homozygous for geotactic response. Thus, using selection techniques, an association between the third chromosome karyotypes and the behavioural trait geotaxis has been found in D. pseudoobscura. It is worth recalling here the associations of karyotype with mating speed and duration of copulation discussed previously.

In these geotaxis experiments, some of the response to selection has been localized to specific chromosomes. In theory it is possible to take this type of analysis substantially further, by using multiply-marked chromosomes. From a back-cross of a female, heterozygous for a recessive multiply-marked chromosome and the selection line, to a male homozygous for the multiply-marked chromosome, recombinants will occur which are recognized by the segregation of marker genes. Certain regions having genetic activity controlling the trait can then be detected, and ultimately genes can be isolated. Thoday (1961) and his colleagues pioneered this approach and have used it with some success in their work on sternopleural chaeta number (see Lee and Parsons 1968 for references to this and other work). With the exception of some work by MacBean (pages 67–70) for duration of copulation, no behavioural work has been reported which proceeds beyond the whole chromosome level, but there are no *a priori* reasons why it should not, and indeed a trait such as geotaxis would seem to be suitable because of the ease with which large numbers of flies can be classified for it.

MacBean and Parsons (1967) selected for long and short durations of copulation in *D. melanogaster*, which gave responses in both directions. Initially, there was a general reduction in mean duration for all lines (Fig. 6.4), and two causes come to mind, namely, inbreeding and adaptation to laboratory conditions. The second hypothesis seems more likely since in other strains set up from the wild (Hosgood and Parsons, unpublished data), there was a 5- to 8-minute reduction between the fourth and fourteenth laboratory generations. In the low-selection line a response was attained by generation 9,

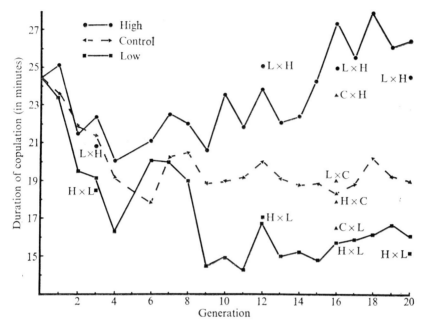

FIG. 6.4. Response to selection for high and low duration of copulation (minutes). At generations 3, 12, 16, and 20 various crosses between selection lines (H and L) and controls were made: in each case the female is written first. The results of the crosses follow the genotypes of the males (see text). (From MacBean and Parsons 1967.)

after which there was little change, and in the high line the major response occurred from generations 13 to 16, leading to over 10 minutes' divergence between the high and low lines at generation 20. Crosses were made between various lines at generations 3, 12, 16, and 20 (Fig. 6.4), and show that, throughout, duration of copulation remains under the control of the male. In the low line a correlated reduction in the time period before sperm was transferred (the pre-transmission period) was found. Because duration of copulation is almost entirely male-determined and so is not determined by an interaction between the sexes, it seemed a suitable trait for which to attempt further genetic analysis.

68 PART I: BEHAVIOUR GENETICS

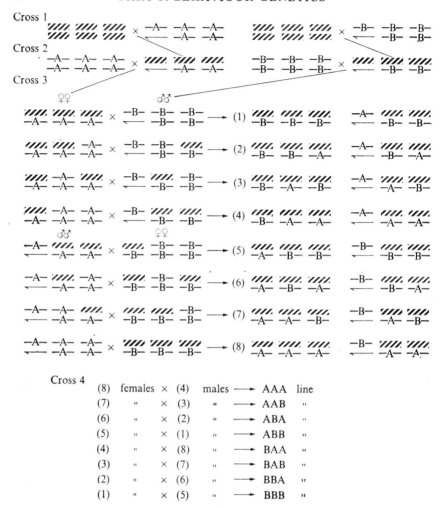

FIG. 6.5. Crossing scheme employed in the production of chromosome substitution lines using the marker stock *Basc*; *Cy ds*^{33k}/*Pm*; *H*/*Sb*. The marker chromosomes are hatched. For the source of the A and B chromosome, see text.

MacBean (1970) (see footnote †) used the procedure of Kearsey and Kojima (1967) for his initial chromosomal analysis. Designating one strain A and another B, and neglecting the Y and the fourth chromosomes, eight true-breeding combinations of the three major chromosomes can be built up from the two base strains A and B. The eight true-breeding lines are AAA, AAB, ABA, ABB, BAA, BAB, BBA, and BBB, where the sequence of letters

† MacBean, I. T. (1970). *The genetic control of quantitative characters in* Drosophila. (Ph.D thesis, La Trobe University).

corresponds to the source of the X, II and III chromsomes. Synthesis of these lines involved the multiple inversion stock *Basc*; *Cy ds*^{33k}/*Pm*; *H*/*Sb*; the crossing procedure is outlined in Fig. 6.5. By crossing the eight true-breeding lines above in various combinations, all the possible twenty-seven female and eighteen male mating combinations can be obtained (Table 6.2). In contrast with the initial chromosomal analysis described for geotaxis, this analysis gives comparisons between the two strains under test, rather than in relation to a tester stock, and so seems to be a preferable method.

TABLE 6.2

Matings between substitution lines

Females	\multicolumn{8}{c}{Males}
	AAA	AAB	ABA	ABB	BAA	BAB	BBA	BBB
AAA	AAA				HAA (A)	HAH (A)	HHA (A)	
AAB	AAH	AAB		AHB				
ABA	AHA		ABA				HBA (A)	
ABB	AHH		ABH	ABB	HHH (A)			
BAA					BAA	BAH		BHH
BAB		HAB (B)				BAB		BHB
BBA					BHA		BBA	BBH
BBB		HHB (B)	HBH (B)	HBB (B)				BBB

The letters in parentheses denote the sources of the Y chromosome. H denotes a chromosomal heterozygote between A and B, while the letters A and B without parentheses denote chromosomal homozygotes.

After Kearsey and Kojima 1967

Analyses of variance are given in Table 6.3 for the high-duration-control contrast, and for the control-low-duration contrast.

For the high-control comparison there were significant effects associated with all chromosomes and for the interaction between the two autosomes. For the control-low comparison, the main effect was associated with the second chromosome, which showed a smaller but significant interaction with the X chromosome. Although selection had been equally effective in both directions, the low line only was analysed in further detail, mainly because most of the response could be assigned to chromosome II. Further analysis showed that much of the genetic difference for reduced duration of copulation in the low line relative to control is localized in the right arm of

TABLE 6.3

Analyses of variance of the chromosomal effects for duration of copulation for the high-control and the control-low comparisons†

Source of variation	d.f.	High-control F	Control-low F
X	1	9·76‡	0·04
II	1	24·53§	24·06§
III	1	74·09§	3·02
X×II	1	0·78	4·19"
X×III	1	0·09	0·18
II×III	1	21·15§	0·01
X×II×III	1	1·25	0·01
Lines	7	18·81§	4·50§
Replicates	1	3·34	48·87§
Lines × replicates	7	1·28	3·01‡
Error	176		

Values of significant chromosome effects (P<0·05)				
	X	+1·10 minutes		
	II	+1·75 minutes	II	−1·76 minutes
	III	+3·04 minutes		
	II×III	+1·63 minutes	X×II	+0·74 minutes

†Based on 2 replicates of 12 and 9 observations for the 8 lines for the high-control and control-low contrasts respectively.
‡$P<0·01$, §$P<0·001$, "$<P0·05$.
After MacBean, I. T. (1970). *The genetic control of quantitative characters in Drosophila*. (Ph.D. thesis, La Trobe University)

chromosome II. Tests also revealed that the second chromosome effect involved the pre-transmission period.

Therefore for two traits, geotaxis and duration of copulation, genetic activity has been localized to the chromosomal level, and for the latter a first attempt at location within chromosomes has been made. There is no reason why more detailed studies would not be worth carrying out.

3. Examples (mainly behavioural analysis—sexual behaviour)

In *D. melanogaster* Manning (1961) selected for fast and slow mating speeds based on fifty pairs of virgin flies, aged 4–6 days, mated together in a mating chamber. The ten fastest and ten slowest pairs were taken by an aspirator to initiate fast and slow selection lines. The response to selection was almost immediate, and two fast and two slow lines were formed, while a control line was maintained during part of the experiment. The rapid response to selection would be predicted from the heritabilities for mating speed given in Chapter 5. After 25 generations the mean mating speed was about 3 minutes in the fast lines and 80 minutes in the slow lines. The divergence between the lines at generation 18 is shown in Fig. 6.6. Considerable variations in speed due to environmental fluctuations occurred during selection, but the fluctuations were generally similar in all lines for a given generation. Little genetic analysis was carried out on the selection lines except that an

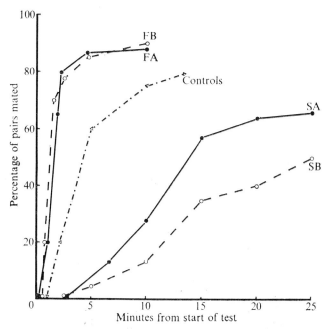

FIG. 6.6. Mating speeds in *D. melanogaster* for two lines selected for fast speeds (FA, FB), two for slow speeds (SA, SB), and controls, at the 18th generation of selection. (After Manning 1961.)

approximate heritability of 0·30 was computed from the rate at which the selection lines diverged during the first few generations.

Manning (1961), however, analysed the behavioural consequences of selection in some detail. Hybridizing the fast and slow lines in both directions gave intermediate F_1 mating speeds, while inter-crossing the two fast and two slow lines themselves in both directions gave fast and slow speeds respectively. These results indicate that both sexes were affected by selection. Confirmation of this came from testing mating speeds against an unselected stock of flies, when both sexes of the selected lines gave altered mating speeds in the expected directions. Activity differences between lines were measured by admitting flies to an arena where the number of squares entered by a fly in a given time period was scored. The slow line exhibited much more activity of this type, which Manning called 'general activity', than did the fast lines. Experiments using unselected females with selected males showed that the lag before courtship was much smaller for the fast than the slow lines; similarly, the frequency of licking was higher in the fast than in the slow lines. Thus, the fast lines have a high level of 'sexual activity' and a low level of 'general activity', and the slow lines a low level of 'sexual activity' and a high level of 'general activity'. Under natural conditions these two components are presumably at an optimum, as over-responsiveness in either direction would be undesirable.

A difficulty for this trait is that selection operates on both sexes and there could be rather different genes controlling the response in the two sexes. Manning (1963) attempted to look into this further by selecting for mating speed based on the behaviour of one sex only. There was no response to selection for fast-mating males or slow-mating females, and a fast-mating female line was not set up. There was, however, a response in male lines selected for slow mating. In these lines, the mating speed of females was unaffected in early generations but somewhat reduced in later generations. Behaviourally it was found that both sexes showed lower general activity, and that the males showed reduced courtship activity, which contrasts with the earlier experiments. Manning was unable, however, to come to any definite conclusions concerning possible differences between sexes in the genetic control of mating behaviour.

It seems that genes affecting general activity and sexual activity must be wholly or partly controlled by separate genetic systems, which can be changed independently. Ewing (1963) selected for spontaneous activity and found that his inactive lines displayed greater sexual activity, as would be expected from Manning's observations. However, Ewing's technique involved placing fifty flies of one sex into the first tube of a line of interconnected tubes and selecting the first ten (active) and the last ten (inactive) to reach the opposite end. This procedure separated flies that went through rapidly from slow ones, but when the two lines so formed were tested by introducing single flies into Manning's arena there was no significant difference between them, so that the two types of behaviour in fact differ. Manning was measuring spontaneous activity, and Ewing was concerned more with dispersal activity or reactivity of flies with each other.

In a more physiological study of the control of receptivity, Manning (1966, 1967) found female acceptance of a courting male to depend on two processes. The first determines whether a female is accessible to the courtship of males. Young females are unresponsive for some 36 hours after eclosion, and then quite suddenly they become receptive and accept a male after a few minutes of courtship. The evidence suggests that this rapid behavioural change of receptivity being switched on, occurs when the concentration of circulating juvenile hormone rises with the reactivation of the corpus allatum after eclosion. Furthermore, the corpus allatum and the ovaries show a growth cycle parallel to that of receptivity. The second process can be called 'courtship summation', and involves the summation of all the heterogeneous stimuli provided by the courting male, which induces a female to allow him to mount once a critical level of stimulation has been reached. The evidence advanced indicates that the two processes are distinct, and that the change from the unreceptive state to the receptive is an all-or-nothing process; that is, females are completely unresponsive to courtship or they accept within the normal time range for receptive females (about 95 per cent of females accepting males within 15 minutes). There is no evidence that females become gradually more receptive by requiring less and less courtship before accepting. Virgin females remain receptive for many days but after the first week of adult life an increasing proportion become unreceptive, and the switch-off seems to be a rapid all-or-nothing event like the switch-on. Old females which have mated and used up their sperm are more often receptive than are virgins of the same age. It is suggested that this is because their corpora allata are more active, and the juvenile hormone concentration is kept above the critical level for longer.

Manning (1968b) selected successfully for slow mating speed in *D. simulans*, in which the behaviour of males was not affected but in which there were marked effects in females (Fig. 6.7), in contrast with the mating-speed selection experiments in *D. melanogaster* where both sexes were affected. Most of the slow-mating line females failed to become receptive on the second day from eclosion as do normal flies. The females in fact performed the most vigorous repelling movements, by extruding the ovipositor and twisting or lifting their abdomen beyond the reach of the courting male. These are movements normally shown only by elderly virgins who have become unreceptive, or by fertilized females whose receptivity is inhibited by the presence of stored sperm in their seminal receptacles. However, the females had normal ovarian growth, and their corpus allatum complex, when implanted into normal hosts, was capable of producing precocious receptivity. The experiments suggest that the females have a normal supply of juvenile hormone, and that the genetic change involves one or more links in the chain of neural 'target organs' on which the juvenile hormone acts, and thus the females do not become receptive. As Manning (1968b) pointed out, this situation shows comparability to that in some mammals, such as guinea-pigs.

Kessler (1968, 1969) selected for fast and slow mating speed in *D. pseudoobscura*, based on an inter-cross between three wild-type strains (from British Columbia, California, and Guatemala), using a technique essentially similar to Manning's (for heritabilities see §5.2). After the 12th generation of selection, tests were made on all possible combinations of fast, slow, and control lines as a 3×3 diallel cross in which observations were carried out for 30 minutes using fifty pairs in one container, copulating pairs being removed with an aspirator. It was found that slow-mating females reduced mating whenever they were involved, and that fast-mating females were not significantly faster than the controls. Fast-mating males speeded up all matings in which they were involved, but slow-mating males were not significantly different from the controls. An analysis of variance showed females to

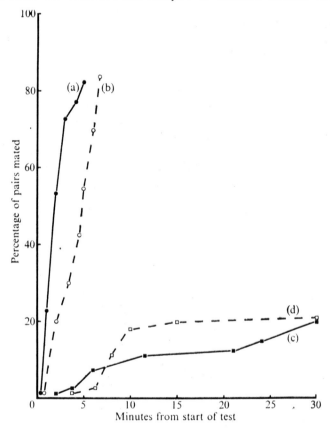

FIG. 6.7. Mating speeds in *D. simulans* for (a) selected ♂♂ × control ♀♀, (b) control ♂♂ × control ♀♀, (c) selected ♂♂ × selected ♀♀, (d) control ♂♂ × selected ♀♀. (After Manning 1968*b*.)

account for more of the total variance than did males. This contrasts with data of Kaul and Parsons (1965) for *ST/ST, ST/CH,* and *CH/CH* karyotypes referred to in Chapter 3, where male determination was very strong, but it should be noted, first, that Kessler was dealing, not with known karyotypes but with selection lines, and, second, that he was dealing with a mating chamber with fifty pairs of flies, while Kaul and Parsons were dealing with single-pair matings.

4. Examples (mainly behavioural analysis—non-sexual traits)

Various other traits have been subjected to directional selection: thus Connolly (1966) selected for locomotor activity in an open-field apparatus made from white Perspex fitted with a colourless Perspex lid. The apparatus measured $10 \times 10 \times 0.5$ cm., and the outer surface of the lid was graduated into 1-cm. squares. Flies were observed for one minute after allowing them 1·5 minutes to recover from the tapping involved in allowing them to enter through a funnel inserted in a small hole in the side of the chamber. An activity score was obtained by counting the number of squares an insect crossed each minute. Quite rapid responses to selection were found in both directions, as would be expected from an estimate of heritability in the narrow sense (h_N^2) of 0.51 ± 0.10 obtained from the regression of offspring on midparent (see §5.4). In order to establish that the response to selection represents a real change in activity and is not an artefact of the measuring technique, three other kinds of apparatus were used and all gave significant differences in agreement with the responses to selection found. No genetic analyses were undertaken, but clearly the results obtained cannot be regarded as unexpected in view of Manning's results.

Grant and Mettler (1969) selected for 'escape' behaviour in *D. melanogaster*. When disturbed by an external stimulus, flies tend to become quite excited and demonstrate an escape response characterized by what superficially appears to be positive phototaxis and negative geotaxis. This was assessed by an appropriately designed maze. Responses to directional selection were quite marked for both high and low maze activity, and a small but significant degree of mating discrimination was found as a correlated response to selection. The same trait was also subjected to stabilizing selection which led (as expected) to a slight decrease in variation in one population, but not in another. Disruptive selection led to an increase in variation compared with control populations, as would be expected, but no isolation between extremes was found.

Hirsch and Boudreau (1958) studied phototaxis in *D. melanogaster* in an apparatus consisting of a Y-maze, one arm of which was exposed to light during tests. Quite rapid responses to selection were found for both positive and negative phototaxis, which is reasonable as a heritability in the broad sense was estimated in excess of 0·5. Phototaxis has been studied with mazes having a similar design to those used for geotaxis, and responses to selection have been found in *D. melanogaster* (Hadler 1964), and in *D. pseudoobscura* (Dobzhansky and Spassky 1969). In both cases positive and negative selection lines were rapidly established. On relaxation of selection in *D. pseudoobscura*, convergence occurred almost as rapidly as the divergence under selection, showing that the average phototaxis of natural populations is a trait subject to genetic homeostasis (Dobzhansky and Spassky in fact made a similar observation for geotaxis, as discussed in §6.3). Spassky and Dobzhansky (1967) found that geographic strains of *D. pseudoobscura* and *D. persimilis* respond differentially for phototaxis, showing the great store of variability available in natural populations for phototaxis. In *D. melanogaster* Médioni (1962) found variations between wild strains collected at different localities in the northern hemisphere, such that greater positive phototaxis was shown in more northern flies. These variations are presumably under genetic control, but their significance is unknown. Phototaxis illustrates a rather subtle environmental problem, in that under the usual conditions of handling in the laboratory *D. pseudoobscura* shows positive phototaxis. Pittendrigh (1958), however, found them to be negatively phototactic in contrast to general observations. Lewontin (1959) carried out a series of experiments showing that *D. pseudoobscura* is negatively phototactic under conditions of low excitement, but if flies are forced to walk rapidly, or fly, they lose their negative phototaxis and become strongly attracted to light. Hadler (1964), in fact, listed numerous environmental variables affecting phototaxis in addition to those just mentioned—namely, temperature, time of day at time of test, time since anaesthesia, rearing conditions, time since feeding, energy and wavelength of light, state of darkness adaptation, number of observations or trials per individual, age, and sex. A phototactic response is therefore a product of a particular stimulus-environment, as well as the genotype. It is quite clear that any form of accurate genetic analysis must be based on defined environments, as has been repeatedly stressed.

The other complication is the method used in studying phototaxis, and three different experimental designs have been employed (see Hadler 1964): measurement of the rate at which flies approach a light source at the far end of a tube (e.g., Carpenter 1905, Scott 1943); recording, after a specific period, of the distribution of flies in a field with a directed light source (e.g., Carpenter 1905, Wolken, Mellon, and Contis 1957); mazes, as already described for geotaxis. Hadler considers that one of the major difficulties in comparing phototaxis from different laboratories is due to such differences in experimental technique, which may be measuring slightly different things—for example, the first method confounds phototaxis with photokinesis.

Optometer response in *D. melanogaster* is another trait for which evidence of genetic determination has come from selection experiments (Siegel 1967). The optometer response to an illuminated moving striped plate was measured, and each fly was given ten opportunities to respond. Optometer scores ranged from zero (no response) to ten. Selective breeding procedures based on low, middle, and high scores were instituted. This led to the development of three strains differing with respect to optomotor response (see also §3.5 for differences between various mutants).

Becker (1970) has made a start on the genetics of chemotaxis of *D. melanogaster* with a Y-unit maze of the type used for geotaxis and phototaxis.

Selection over 12 generations yielded two lines insensitive to insect repellants, and crosses indicated that the genes responsible for insensitivity were at least in part dominant. As compared with geotaxis and phototaxis, chemotaxis seems to have the likely advantage of being able to relate the stimulating molecule to the specificity of the receptor, and further attempts to select genetic variants for chemotaxis will be awaited with interest.

A number of other traits may be amenable to genetic analysis, especially with selection techniques, but as yet have been discussed mainly from the behavioural point of view. One such trait is the preening or cleaning behaviour which has been described in terms of a number of separate behavioural elements (Connolly 1968). The various preening movements serve to keep the insect clean, and free the sensory surfaces from contaminants. The presence of other flies increased the amount of preening behaviour, and this was shown not to be a function of physical contacts between the flies. Further descriptions of this type of behaviour are given by Szebenyi (1969), whose approach is analogous to Bastock's (1956) detailed study of mating behaviour of yellow and wild-type flies since he analysed preening or cleaning behaviour into a series of components. Szebenyi thus considers preening to be a rich source for behaviour-genetic analysis. Hay (1972*a–c*) in fact has indicated this for both preening and activity using biometrical methods (see §5.3).

7

DEVIATIONS FROM RANDOM MATING

1. Introduction

IN Chapter 3 various mating-choice experiments were described indicating non-random mating due to variations in sexual vigour (sexual selection); in this chapter other forms of non-random mating will be considered. Any form of non-random mating has some significance for the population in that it represents a deviation from the Hardy-Weinberg Law on which much of population genetics theory is based.

Two other main classes of deviation will be considered. First, there is *density-dependent mating* where the frequencies of matings depend on the frequencies of genotypes; in particular, the rare genotypes tend to be favoured at the expense of the common. Second, there is sexual isolation, where homogamic matings exceed heterogamic matings, which, as pointed out in Chapter 3, is not the same thing as sexual selection. Within a population or species, the term *positive assortative mating* is often used to refer to a situation where there is a general tendency for like phenotypes to mate. Conversely, the term *negative assortative mating* is used to refer to a rarer situation where there is a general tendency for unlike phenotypes to mate. The terms positive assortative mating and sexual isolation are in a sense synonymous, although sexual isolation is used more widely and is usual when discussing mating behaviour between races and species, whereas positive assortative mating is mainly restricted to phenomena within a given population.

2. Density-dependent mating

Petit (1951, 1954, 1958) carried out fairly extensive multiple-choice experiments in *D. melanogaster* and showed that the influence of the genotype of the females was unimportant compared with that of the males in her experiments. For the sex-linked gene Bar, she found that Bar males were always at a disadvantage compared with wild-type males, but that this disadvantage was most pronounced when Bar was most frequent. For white and wild-type males, the white males were at a disadvantage when 40–80 per cent of the males in the population were white, but outside these limits when white males were rare or abundant they had the advantage. Thus mating success depends on the *proportion* as well as on the nature of the competing genotypes; and there is some suggestion of rare genotypes being favoured.

FIG. 7.1. Photograph of a double observation chamber for the observation of *Drosophila* matings. The matings taking place in the larger chamber are recorded. The smaller compartment contains the flies producing the contaminating odours. A blower is attached by a rubber tube to a nozzle in the lower right corner.

More recently, a number of reports have appeared suggesting that density-dependent effects associated with mating occur quite frequently (see Petit and Ehrman 1969), such that the genotype present in a minority frequency mates more frequently than would be expected on a random mating hypothesis. One of these reports was due to Spiess (in Ehrman 1966), and involved flies homozygous for two autosomal inversions Klamath (*KL*) and Whitney (*WT*) in *D. persimilis*, where the proportions of *WT* and *KL* males were varied and showed a clear advantage to the minority homokaryotype. Ehrman (1966, 1967, 1968) and Ehrman, Spassky, Pavlovsky, and Dobzhansky (1965) (and also see Petit and Ehrman 1969 for review) have shown the same in *D. pseudoobscura* using Elens-Wattiaux observation chambers (see Fig. 3.3, page 21) for strains with different karyotypes and geographic origins, mutant versus wild types, positively and negatively selected geotactic lines, and flies raised at different temperatures. Frequency-dependence seems mainly to depend on the frequency of males. Similar observations (Ehrman and Petit 1968) have been made in other species such as *D. tropicalis, D. willistoni*, and *D. equinoxalis* for minority males from different geographic strains, and to a lesser extent the same was true of minority females (see also Spiess 1968*b*, and Spiess, L. D. and Spiess, E. B. 1969).

Thus, in the last few years the generality of the phenomenon whereby rare genotypes as males in various *Drosophila* species are favoured in mating has become accepted, and it is reasonable to inquire into possible mechanisms. Ehrman (1969) has looked into the sensory basis of mate discrimination. She regards olfaction as the cue, since in cages (see Fig. 7.1) in which males are rare, the minority advantage disappears if currents of air derived from a section of the cage containing courting and copulating couples of the rare type flow over the flies. The air current apparently delivers an olfactory cue to the observation chambers which obscures the olfactory rarity of the rare type. Or, put in another way, when odours of the rare male type are present, *Drosophila* females can no longer discern the rarity of that particular group of courting males, and therefore no longer accept them more frequently than expected on a random basis. Recent work of Shorey and Bartell (1970) is of interest in this connexion, since they have evidence for a volatile sex pheromone produced by the females of *D. melanogaster* which stimulates males to courtship. Their conclusion was that the female sex pheromone increases the probability that a male will initiate courtship with a nearby female. Male courtship behaviour is stimulated by the odour released from other males as well as other females; nevertheless, the odour of males appears to be less than one-tenth as effective as is that of females. It seems clear that the influence of hormones, pheromones, and other physiological processes will be increasingly studied in the next few years for mating behaviour and other behavioural traits.

From the population point of view, the advantage of the rare male would be expected to lead to an increase in its frequency, provided that there are no other selective forces acting against it. As the rare type of male increases, the mating advantage would slowly diminish, and at the stage when it disappears a selective equilibrium could well be expected. It seems likely that a number of gene and chromosomal polymorphisms in *Drosophila* are maintained by such frequency-dependent balancing selection, based on mating behaviour. These are polymorphisms for which minimal fitness differentials between the component genotypes are expected at equilibrium, unlike the heterozygote advantage model (see Appendix to Chapter 4), and therefore represent a way of maintaining a high level of genetic variability, without associated fitness differentials. This may be of some considerable evolutionary significance, since it has been argued that there is a limit to the number of polymorphisms a population can maintain under the classic heterozygote fitness advantage model (see Chapter 13 for further discussion).

In most of the work so far cited, various techniques of direct observation of the numbers of matings taking place among flies confined to the small space of an observation chamber were used. Ehrman (1970*a*) carried out some experiments where *D. pseudoobscura* were allowed to mate in mass cultures, and the proportions of two types in each generation were determined according to their mating success in the previous generation. From initial values of 80 per cent for the common and 20 per cent for the rare type, the frequencies converged to approximate equality because the rare males were favoured as mates. However, when the formerly rare type increased in frequency, it lost its mating advantage and a balanced equilibrium was eventually attained. Even more significant are some experiments carried out by Ehrman (1970*b*) in a room with a volume somewhat in excess of 2655 cubic feet using *D. pseudoobscura* homozygous for the autosomal recessive orange-eyed mutation, *or*, and for the wild-type phenotype. Based on two experiments in which 2000 flies were released in the ratio 8 *or* : 2+ in the first experiment, and the reverse in the second, an advantage of the rare genotypes was found, especially in the first. This is the closest approximation so far to natural populations, and suggests that if the phenomenon is widespread in natural populations it may play a considerable role in evolution.

3. Sexual isolation

Sexual isolation leads to a modification of the pattern of random mating within or between species; different strains within species of *Drosophila* are considered first. In *D. pseudoobscura* Anderson and Ehrman (1969) studied five geographically distinct populations using Elens-Wattiaux mating chambers and the multiple-choice method. Ten virgin females and males were used (Table 7.1) and the results indicated that only in one case, the Berkeley × Okanagan cross, is there evidence for sexual selection. This was due to the

TABLE 7.1

Mating preferences in crosses between geographic populations of Drosophila pseudoobscura

Cross Strain A × Strain B	Number of chambers run	Number of matings	Numbers of each type of mating				χ_3^2 for random mating
			A♀×A♂	A♀×B♂	B♀×A♂	B♀×B♂	
Berkeley × Okanagan	12	222	60	50	72	40	10·14†
Berkeley × Austin	8	160	37	43	42	38	0·65
Berkeley × Hayden	2	28	7	7	5	9	1·14
Berkeley × Sonora	8	103	23	22	28	30	1·74
Okanagan × Austin	8	125	27	33	33	32	0·79
Okanagan × Hayden	7	51	14	14	10	13	0·84
Okanagan × Sonora	7	114	26	29	32	27	0·74
Austin × Hayden	8	103	21	26	30	26	1·58
Austin × Sonora	8	113	36	28	27	22	3·57

† Significant at 0·05 level.

After Anderson and Ehrman 1969

significantly increased activity of Berkeley males. Calculation of isolation indices showed no evidence for non-random mating. In *D. willistoni* slight mating preferences have been shown in male-choice experiments between Brazilian and Guatemalan strains (Dobzhansky and Mayr 1944), and in *D. prosaltans* striking differences were found between Brazilian, Guatemalan, and Mexican strains (Dobzhansky and Streisinger 1944). In these two species most of the mating preferences were one-sided, as a greater proportion of females of one than of the other strain were inseminated by males of both strains—that is, in these cases there is sexual selection rather than sexual isolation.

However, in *D. sturtevanti* male-choice experiments show two-sided mating preferences for homogamic matings, so that there is a tendency towards sexual isolation (Dobzhansky 1944). The strains came from Tamazunchale in Mexico, Quiriguá in Guatemala, Belém in northern Brazil, Rio de Janeiro, and Bertioga in the state of São Paulo, Brazil. In all the possible crosses, positive isolation indices were obtained, showing that the males inseminate more females of their own strain than those of other strains. The highest indices came from the two geographically most remote localities (Tamazunchale, Mexico, and Bertioga, Brazil), although some localities nearly as remote (Tamazunchale and Rio de Janeiro) failed to show statistically significant isolation indices. Strains from geographically close localities either showed no isolation (Rio de Janeiro and Bertioga) or considerable isolation (Tamazunchale and Quiriguá), and strains showing little or no isolation from each other occasionally behaved very differently with respect to other strains. However, in spite of the fact that correlation with geographical isolation is far from clear, the strains of *D. sturtevanti* from different regions do show incipient sexual isolation because of the excess of homogamic matings. More recently Miller and Westphal (1967) have reported reproductively isolated subdivisions in the north American species *D. athabasca*, although in the laboratory the isolation is not complete.

The hybrids between two strains each possessing a genotype well-adapted to its own environment are likely to be ill-balanced and unfit for survival. This will reduce the effective gene exchange between populations, so that a reduction of gene exchange may be favoured by natural selection. Any genetic variants reducing the rate of exchange will be favoured until a complete cessation of gene exchange is obtained. This may be aided by any form of isolating mechanism, including sexual isolation, which may evolve during divergence. Dobzhansky (1944) regards the one-sided mating preferences found regularly for mutant genes and for populations within certain species as by-products of physiological (and hence behavioural) differences. He considers that sexual isolation may develop from one-sided mating preferences by the co-ordinating action of natural selection.

D. paulistorum is a taxon which contains an extraordinary complex of geographic races or incipient species. Six are near the status of reproductively isolated but morphologically indistinguishable species. They are the Centro-American, Orinocan, Amazonian, Andean-South Brazilian, Interior, and Guianan races (Carmody, Collazo, Dobzhansky, Ehrman, Jaffrey, Kimball, Obrebski, Silagi, Tidwell, and Ullrich 1962; Perez-Salas, Richmond, Pavlovsky, Kastritsis, Ehrman, and Dobzhansky 1970). These are mainly but not always allopatric (that is, occurring in different geographic localities), and when placed together females of one and males of another race exhibit pronounced and, in some cases, nearly complete sexual isolation. If hybrids are produced, the females are fertile and the males sterile. Observations of the courtship behaviour patterns (Kessler 1962) show both qualitative and quantitative differences between strains, the courtship pattern of the Guianan race being conspicuously different from the others, which tends to promote the reproductive isolation of the Guianan race from the others as shown by a low rate of courtship of interracial females. Kastritsis and Dobzhansky (1967) in fact regard the Guianan race as a full species, *D. pavlovskiana*.

A further race, called the Transitional race, occurs in Colombia. All strains of this race can be crossed and will produce fertile hybrids with at least one of the other races (Dobzhansky and Spassky 1959), with the proviso that more information is required on the recently described Interior race. Dobzhansky, Pavlovsky, and Ehrman (1969) regard the Transitional race as a relic of the ancestral population from which the other races have differentiated. Because of the Transitional race, *D. paulistorum* can perhaps be regarded as a single species, although this is a matter of opinion. It therefore has a single but extremely dissected and differentiated gene pool, and represents a good example of a situation where it is difficult to decide whether there are one or several species.

Carmody *et al.* (1962) scored by female dissection tests, a series of sixty-seven male-choice experiments where all flies belonged to the same race but usually to strains of different geographic origin. About half of the isolation indices were significant, at least at the 5 per cent level, which suggests that random mating is by no means a general rule in intraracial crosses. Positive indices were more common than negative indices, so that even for geographic strains of the same race homogamic matings are more frequent than heterogamic ones. For interracial crosses positive isolation indices were quite general, although there were variations between races. It is noteworthy that the Transitional race is transitional with regard to sexual preference, since the average isolation index in crosses of other races with the Transitional race was $+0.65$, while in crosses between other races it was $+0.87$ (see Table 7.2).

One final observation (Ehrman 1965) of interest is based on the multiple-choice method using the Elens-Wattiaux mating chambers and scoring by direct observation. Joint isolation coefficients were computed for given pairs

TABLE 7.2

Mean isolation indices in crosses between different races of D. paulistorum

Cross	Isolation index†
Transitional × Centro-American	+0·675
× Amazonian	+0·577
× Andean	+0·458
× Orinocan	+0·824
× Guianan	+0·775
Average with Transitional	+0·650
Amazonian × Andean	+0·945
Amazonian × Guianan	+0·860
Andean × Guianan	+0·895
Centro-American × Amazonian	+1·000
× Andean	+0·777
× Orinocan	+0·785
× Guianan	+1·000
Orinocan × Amazonian	+0·883
× Andean	+0·837
× Guianan	+0·875
Average without Transitional	+0·874

† Means of indices derived from a series of male-choice experiments: so far as possible all insemination frequencies were kept close to 50 per cent, so that indices from different experiments could be compared.

After Carmody *et al.* 1962

of races which have been found to occur both sympatrically and allopatrically. In allopatric crosses the average isolation coefficient was +0·67, and in sympatric crosses +0·85. Thus, pairs of races occurring sympatrically exhibit a greater degree of sexual isolation than the same pairs occurring allopatrically, or races coexisting geographically tend to be reproductively more isolated than those that do not, which is reasonable as the production of large numbers of hybrids would be very inefficient.

So far two isolating mechanisms, sexual isolation and hybrid sterility, have been mentioned. Experimental work has shown that sexual isolation is probably determined by polygenic differences between the races, while hybrid sterility involves a remarkable predetermination of the egg cytoplasm by the genotype present in the egg before meiosis and fertilization (Ehrman 1960, 1961). An interesting form of hybrid sterility which was recently reported is the induction of sterility in non-hybrid males by an 'infective agent' (Williamson and Ehrman 1967). Their findings, together with experiments reported by Dobzhansky and Pavlovsky (1967) on strains in nature, indicate the possible importance of this form of species formation. The hypothesis is that the process of speciation may be initiated by the establishment of new symbiotic relationships between *Drosophila* and a virus or other microorganism.

In conclusion, *D. paulistorum* makes up a biologically interesting situation as it contains populations at all stages of isolation ranging from apparent identity to almost complete separation. The variations observed presumably have arisen by natural selection occurring as a result of geographic separation and the accumulation of genetic differences in the course of adjustment to different environments. The second stage involves selection for stronger isolation should the populations become sympatric, which might eventually become complete, at which stage the strains could be regarded as species. The species of *Drosophila* generally, therefore show how there is a gradation from species showing mainly random mating between geographic strains to species showing mainly sexual selection, and to those showing sexual isolation which increases in degree as the dividing line between species is approached. Geographical isolation may, though not necessarily, lead to sexual isolation. Broadly speaking, there seems to be a correlation between geographical and sexual isolation, but there are numerous exceptions. It should finally be pointed out, following Anderson and Ehrman (1969), that the results could be biased, because negative results from such studies are less likely to be published, and because significant isolation is more likely to attract the attention of a reviewer. In this sense, the results of Anderson and Ehrman (1969) on *D. pseudoobscura* are particularly valuable.

The expectation is for complete sexual isolation between species, since all matings would be expected to be homogamic giving complete sexual isolation. However, it is pertinent to comment briefly on the sexual behaviour of closely related species. In general (Spieth 1952, 1958), closely related species of *Drosophila* have the same basic courtship and mating behaviour. The differences which apparently prevent the successful synchronization of the behaviour patterns of the two sexes in interspecific combinations are quantitative rather than qualitative. However, as species diverge phylogenetically as determined by various criteria such as structure, ecology, distribution, etc., the observable mating behaviour diverges more obviously (see also Brown 1965). As Spieth points out, exceptions may occur to these rules, but they form reasonable generalizations.

Spieth (1958) mentions some pairs of species which appear to have qualitatively similar mating behaviour and live in the same area. That is, they are sympatric, and yet at the same time they maintain their identity and so represent separate species. We wish to know how these populations are isolated, and whether ethological isolating mechanisms are relevant. When such pairs of species are morphologically almost indistinguishable, they are referred to as sibling species. Such pairs of species that have been studied in detail show a complex of isolating mechanisms, some behavioural and some ecological. Among these can be cited (1) *D. pseudoobscura* and *D. persimilis*, (2) *D. melanogaster* and *D. simulans*, and (3) *D. gaucha* and *D. pavani*. Mating is now far from random and it is also increasingly difficult to separate behavioural from ecological mechanisms (see Chapter 15 for further discussion).

4. Experimental modification of the degree of sexual selection and isolation within species

The degree of sexual isolation is controlled by natural selection and so is an evolutionary process. Therefore it would be expected that the degree of sexual isolation could be modified by artificial selection in the laboratory. Thus, in *D. melanogaster* Wallace (1954) allowed straw and sepia virgin females and males to interbreed, and in subsequent generations kept the mutant flies and discarded the hybrids, so artificially selecting for homogamic matings. Tests during the seventy-third generation revealed a strong tendency for homogamic sepia matings. Similarly Knight, Robertson, and Waddington (1956) began with a population containing ebony and vestigial flies and each generation discarded the hybrids. After from thirty to thirty-five generations the proportion of hybrids was reduced substantially, probably due to changes of female behaviour such that the females became less willing to cross-breed after selection. Hoenigsberg, Chejne, and Hortobagji-German (1966) selected against heterogamic matings in *D. melanogaster* using ebony and dumpy homozygotes mating with themselves as the homogamic matings. The offspring of heterogamic matings, which were wild-type, were selected out before they could take part in mating. Using the female-choice method, homogamy was shown to increase quite rapidly in frequency such that, by the twenty-first generation of selection, incipient sexual isolation was prevalent in the eleven lines tested. In these various experiments, selection is presumably acting to alter the choice between individuals in courtship; the changes were quite rapid, especially in the last experiment cited. Experiments showing that sexual isolation *between* certain species can be modified in the laboratory is discussed in Chapter 15.

Koref-Santibañez and Waddington (1958) searched for mating preferences between strains of *Drosophila melanogaster* that had been genetically isolated for many generations, using male- and female-choice experiments. In four of six inbred strains they found occasional preferences for homogamic matings in male-choice experiments, and in two of the four the same tendency was found in female-choice experiments. They argued that the tendency towards homogamic matings is initially a chance phenomenon, and so suggest that the initial steps towards sexual isolation may sometimes be attributable to chance. In a further eight strains, of which four had been selected for high, and four for low, abdominal chaeta numbers, they found that mating was nearly at random. Mather and Harrison (1949), using the male-choice method, found mating preferences between two inbred strains of *D. melanogaster*, and also between strains selected for abdominal chaeta number which were based on an initial cross between the two inbred strains. These experiments all show that sexual selection and isolation may arise in certain circumstances as a concomitant to genetic divergence.

In *D. pseudoobscura*, Ehrman (1964) studied sexual isolation between six populations set up in cages at the same time by M. Vetukhiv and derived from the same initial population. Of the populations, two were maintained at each of 16°C, 25°C, and 27°C. After four years and five months male-choice experiments of the types $10A\male\male \times 10A\female\female + 10B\female\female$ and $10B\male\male \times 10A\female\female + 10B\female\female$ were carried out, where A and B represent flies from different cages. The choice experiments were carried out at 16°C and 27°C, the two extremes of temperature at which the populations were kept. Some sexual isolation was found in all cases, but it was as pronounced within temperatures as between temperatures (Table 7.3), so that sexual isolation evidently takes

TABLE 7.3

Joint isolation indices from six isolated populations of D. pseudoobscura

Populations crossed	°C	Index
A × B	16	+0·097
	27	+0·209
C × D	16	+0·090
	27	+0·190
E × F	16	+0·052
	27	+0·106
A × C	16	+0·075
	27	+0·014
A × E	16	+0·154
	27	+0·016
C × E	16	+0·011
	27	+0·105

Environments of population: A and B at 16°C, C and D at 25°C, E and F at 27°C. After Ehrman 1964

place between isolated populations kept in similar as well as in different environments. Sexual isolation has arisen in the absence of any selection for isolation and is evidently a by-product of genetic divergence. Thus geographical separation may facilitate the achievement of reproductive isolation, presumably because within each cage there was some reorganization of the gene pool during the period of isolation, such that each cage built up its own unique highly adapted gene complex, with its own behavioural phenotype. In this way sexual isolation could begin to evolve in the wild.

Ehrman's results are of interest in relation to those of Gibson and Thoday (1962) who practised disruptive (diversifying) selection for high and low sternopleural chaeta numbers in *D. melanogaster* in a single population maintaining gene flow between the high and low components. As a result the population split into two discrete sub-populations characterized by high and low chaeta numbers. This occurred within ten generations. Thus the population became bimodal and hence polymorphic under disruptive selection as was predicted by Mather (1955). This is a polymorphism in which the heterozygotes are at a disadvantage, since under disruptive selection the extremes are favoured. Under random mating the heterozygote must be fitter than the two corresponding homozygotes for a balanced polymorphism to occur (see Chapter 4).

Maynard Smith (1962) has shown that a polymorphism in the absence of heterozygote advantage can be maintained under the specialized situation of the two extreme genotypes of the population being adapted to different ecological niches with gene flow between the niches, such that in niche 1 $AA > Aa > aa$ in fitness, and in niche 2 $aa > Aa > AA$. That is, the fitnesses are in the opposite sense in the two niches, a situation formally equivalent to disruptive selection. This, with one or two further restrictions, led Maynard Smith to argue that, except under special circumstances, it is difficult to imagine disruptive selection leading to rapid isolation in the wild. Mayr (1963) arrived at similar conclusions. Maynard Smith further considered that Gibson and Thoday's (1962) experimental results are very difficult to understand unless there is positive assortative mating for sternopleural chaeta number in the base population, or unless there is selection favouring positive assortative mating during the experiment, that is, selection favouring a greater proportion of flies with high chaeta numbers mating together and low chaeta numbers mating together than would be expected under random mating. Indeed, Thoday (1964) in one series of experiments found strong positive assortative mating within the high and low lines for female-choice tests carried out at various generations between the seventh and nineteenth of disruptive selection, giving a joint isolation coefficient of $+0.52$. Thus, the isolation developed by disruptive selection is associated with strong positive assortative mating within the sub-populations.

In the wild, positive assortative mating has not commonly been observed within populations, although it has been found between the various colour forms of the Blue Snow Goose (Cooch and Beardmore 1959) and the Arctic Skua (O'Donald 1959). The best evidence comes from man where positive assortative mating has been found for numerous physical traits such as stature and span, and for many behavioural traits (see Parsons 1967*a* for references). Assuming that these traits are heritable, then assortative mating must be, as Fisher (1930) argued, an agent important in modifying the genetic constitution of a population.

The question then remains as to whether there is positive assortative mating in *Drosophila* for chaeta number in the base population before disruptive selection, or whether assortative mating arises *de novo* during selection. In *D. melanogaster* the possibility of positive assortative mating for sternopleural chaeta number was tested (Parsons 1965b) by placing forty virgin females and males from an out-bred Canton-S strain, aged 3 days, in a mating chamber. As soon as a pair commenced to mate it was sucked into a trap and then stored separately to await scoring for sternopleural chaeta number. Mating pairs were extracted until about one-half of the pairs had mated and then the remaining flies were stored together. Correlation coefficients between mated males and females for sternopleural chaeta number were in the range $0 \cdot 1 - 0 \cdot 2$ and were all significantly >0, so indicating positive assortative mating. For unmated flies, correlation coefficients were calculated by arbitrarily pairing together flies as they were scored, so that the coefficients obtained, which were all close to 0, represent control values, as pairing was approximately at random. Unpublished results give similar coefficients for abdominal chaeta number. These results could be a direct effect of fly size, as sternopleural chaeta number and fly size are directly correlated when fly size is altered by environmental means (Parsons 1961), or perhaps there may be behavioural differences between flies of different sizes leading to minor modifications in courtship behaviour. For example, wing area, which is probably related to fly size, is a factor in determining male sexual success (Ewing 1964). Thus there is positive assortative mating for sternopleural chaeta number in a *Drosophila* population, which may be accentuated by disruptive selection. There is a need to find out to what extent assortative mating occurs in natural populations for other quantitative traits, especially in those cases where the genetic control of the trait can be studied.

PART II

8

GENETIC HETEROGENEITY FOR ENVIRONMENTAL STRESSES

1. Temperature and desiccation

IN *D. melanogaster* Hosgood and Parsons (1968) showed differences between strains set up from single inseminated females collected in the wild for the ability of adult flies to withstand a high-temperature shock (33·5°C for 24 hours for 7-day-old flies), and Parsons (1969) showed the same for the ability of newly hatched larvae to emerge as adults at 30·5°C. These differences were maintained over a number of generations, showing that the differences between strains are likely to be genetic in origin, presumably arising from genetic differences between the founder females. It follows then, in agreement with work on scutellar chaeta number and other morphological traits (Parsons and Hosgood 1967, Parsons 1968), on the behavioural traits, mating speed, and duration of copulation (Hosgood and Parsons 1967*a*), and on resistance to ^{60}Co γ-rays (Parsons, MacBean, and Lee 1969), that the base population must be polymorphic for genes controlling resistance to high temperatures (see Parsons, Hosgood, and Lee 1967). Therefore the strains differ for genes controlling resistance to high temperatures.

There is other evidence in the literature for temperature-sensitive strains (Ogaki and Nakashima-Tanaka 1966) in *Drosophila*. Tantawy and Mallah (1961) found superiority at high temperatures for populations of *D. melanogaster* and *D. simulans* from Uganda, an area of high temperature. The temperatures used by Hosgood and Parsons (1968) are fairly extreme, but they would occur in the locality from which the flies were derived (Leslie Manor, Victoria), which is characterized by relatively hot summers, so it is reasonable to find evidence for some genetic control of the trait (assuming that the flies do not completely avoid high temperatures by behavioural or other means). It is unlikely that such genes would be fixed in the population, because natural selection would not favour them except at limited times during the year. Such polymorphisms would allow a rapid adjustment to environmental changes as they occur.

The data of Tantawy and Mallah (1961) show that the percentage emergence of flies from eggs was higher for *D. simulans* in the range 18°C to 25°C, whereas over the *entire* range studied, 10°C to 31°C, *D. melanogaster* had higher average survival (see Fig. 8.1). Thus *D. simulans* might be expected to outnumber *D. melanogaster* in regions where temperature extremes are not encountered, but in regions where there are rather wide temperature fluctuations *D. melanogaster* might be expected to displace *D. simulans*. Wallace (1968) points out that data in the U.S.A. fit this expectation fairly well, since *D. simulans* is the more common species of the two in the south where the climate is equitable throughout the year, and *D. melanogaster* is

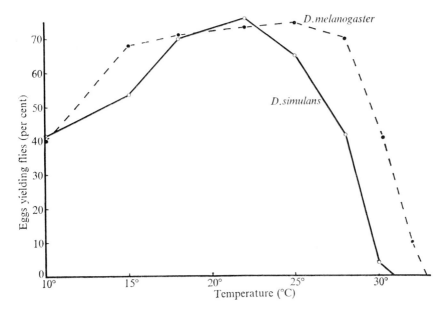

FIG. 8.1. The proportion of eggs yielding adult offspring for *D. melanogaster* and *D. simulans*. (After Tantawy and Mallah 1961.)

more common in the north where winters are severe and wide daily fluctuations in temperature occur. Hosgood and Parsons (1966) set up four strains of *D. melanogaster* and three of *D. simulans*, each derived from single inseminated females collected in the wild, at 29·5°C, 27·5°C, 25°C, 20°C, and 15°C in the laboratory. After five generations, all strains of *D. melanogaster* were living at all temperatures, whereas for *D. simulans* the three strains at 20°C were living but only one at 25°C (which in fact survived for 24 generations). At 29·5°C and 15°C all the *D. simulans* strains had died out by the second generation, and at 27·5°C, by the third generation. Thus *D. simulans* is much more restricted in its tolerance to diverse temperatures than *D. melanogaster*, and this agrees with the United States distribution data reported by Wallace (1968). One problem not answered by these experiments is the effect of fluctuating temperatures, either diurnally or on a longer-term basis, since the length of exposure to extreme temperatures and the amplitude of the temperature variation may be of importance.

Because of the genetic heterogeneity for resistance to temperature, the early observations of Timofeeff-Ressovsky (1940) on 'temperature-races' in *D. funebris* are of considerable interest. The relative viability of *D. funebris* was studied in flies from regions covering the major climatic zones of Europe, northern Africa, and into Asiatic Russia at 15°C, 22°C, and 29°C. It was found that the northern populations were more resistant to cold temperatures, and the southern more resistant to high temperatures. The eastern populations showed resistance to both high and low temperatures, which can be accounted for in terms of the more 'continental' climate of these regions with very low mean winter temperatures and very high mean summer temperatures. Presumably the temperature races differ by the accumulation of appropriate temperature genes into their genome. It would be of interest to study in detail races of *D. persimilis* (see §4.1), which show varying fitnesses at different temperatures in the laboratory, according to whether the flies are derived from high or low altitudes (Spiess and Spiess 1967), the former environment being more 'continental' than the latter. The same comment, of course, applies to a number of other species of *Drosophila*. In *D. flavopilosa*, for inversions on the right arm of the fifth chromosome, Brncic (1968) found certain heterokaryotypes A/Standard to be favoured when larvae develop at 25°C in the laboratory, and others B/Standard at 16°C. The data agree with what is observed in the corresponding natural populations where A/Standard heterokaryotypes are favoured in the summer and B/Standard in the winter (Brncic 1966). As a result of the increase of the A inversion concomitant to the decrease in the B inversion at 25°C, and the reverse at 16°C, heterozygosity for inversions is maintained at a more or less constant level. In their studies of the Hawaiian Drosophilidae, Carson *et al.* (1970) believe temperature to be an important limiting factor. Thus most species are found between 1000 and 5000 feet in elevation, and it is also significant that some of the plant species which serve as larval and adult food sources are abundant in these regions (see Chapter 14).

Another environmental stress of some ecological importance is desiccation. Little work has been done on this trait although Kalmus (1945) and Waddington, Woolf, and Perry (1954) described differences between strains of *D. melanogaster* for preferences for environments with different humidities. Pittendrigh (1958) found that *D. persimilis* lost water by cuticular transpiration more rapidly than *D. pseudoobscura*, and that males lost water more quickly than females. Females seem generally more resistant than males to desiccation in *D. melanogaster* (Kalmus 1941, Perttunen and Salmi 1956). In *D. pseudoobscura*, Heuts (1948) found that the relative humidity during the pupal stage at 25°C differentially affected the carriers of different gene arrangements, thus pupae homozygous for *AR* had higher hatching percentages at low humidities, *CH* homozygotes similarly at high humidities, and *ST* homozygotes were intermediate. Subsequently, Spassky (1951) and Levine (1952) found certain karyotypes to be differentially affected by certain combinations of temperature and humidity. Among other environmental components affecting reaction to humidity is age, in that young *D. melanogaster* of both sexes showed a strong preference for the drier alternative out of 100 per cent and 77 per cent relative humidity. The intensity of the dry reaction gradually decreased during the following days, until after about two weeks the flies were virtually indifferent (Perttunen and Ahonen 1956).

In *D. pseudoobscura*, Thomson (1971) studied desiccation of the *ST/ST*, *ST/CH*, and *CH/CH* karyotypes and showed distinct heterokaryotype advantage (Fig. 8.2) for survival 24–42 hours after the commencement of desiccation. In fact, the only survivors at 42 hours were heterokaryotypes. The

TABLE 8.1

Means and coefficients of variation (c.v.) of live weights of 100 flies of ST/ST, ST/CH, *and* CH/CH *karyotypes in* D. pseudoobscura, *expressed in arbitrary units*

	ST/ST	*ST/CH*	*CH/CH*
Males			
Mean	66·08	73·88	72·43
c.v.	20·07	18·32	25·44
Females			
Mean	80·41	89·51	85·51
c.v.	19·58	17·24	26·78

After Thomson 1971

heterokaryotypes were heavier than the homokaryotypes, suggesting a relationship between ability to survive and body-weight (Table 8.1). Parsons (1970a) found that strains set up from single inseminated females of *D. melanogaster* derived from two populations differed in ability to withstand desiccation, as measured by mortalities after 16 hours in a dry environment, thus indicating genetic variability in wild populations for ability to withstand desiccation. Those strains with high wet and dry weights lost water relatively less rapidly, and had lower mortalities, than strains with lower wet and dry weights. Quite high heritabilities were found for all traits. Levels of resistance to desiccation may have some influence in determining the distribution of the species in the wild. Just as with temperature stress, it is reasonable that there should be genes segregating in the population for resistance

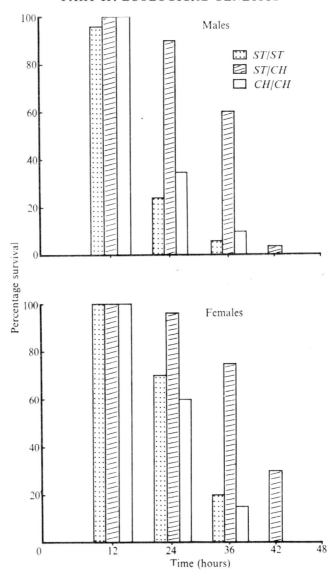

FIG. 8.2. Percentage survivals in males and females for the *ST/ST*, *ST/CH*, and *CH/CH* karyotypes of *D. pseudoobscura* when desiccated. (From Thomson 1971.)

to desiccation, because during parts of the year populations would be exposed to desiccation stresses, when genetic variability to adapt may be desirable. A correlation between the two stresses might often be expected, since in the region of Australia where these strains were collected a correlation between high temperature and low humidity would be common, although in some parts of the world, especially in the tropics, this might not appear likely. However, in the Australian strains tested, a significant correlation was found between the two traits. Clearly, more work would be rewarding especially if, following Levins (1969) and the work just cited, high body-weight can be regarded as an adaptation to desiccation stress. However, Levins (1969) quotes results indicating the likely complexity of these traits in relation to the exact natural environments from which the flies were derived.

Levins (1969) studied acclimation to dry heat (a combination of high temperature and low humidity), and found that a process of physiological acclimation tended especially to occur in broader niched species, such as *D. melanogaster*, and tended not to occur in narrow niched species, such as *D. prosaltans*. In other words, there may be a degree of individual flexibility which allows of immediate adaptation to heat stress, as well as genetic variation as cited above. Levins feels that this work may be extrapolated to the wild, and such acclimation may permit feeding in dry places without immediate desiccation. A further complication may come at the behavioural level, since *D. willistoni* probably tends to avoid stress by behavioural means (Levins 1969). *D. willistoni* is of interest in that it is a moderately broad-niched species showing little acclimation or genetic variation in heat resistance, and hence behavioural avoidance mechanisms may be of importance. If behavioural avoidance is general, then Levins's conclusion of a negative correlation between strong habitat selection and individual (physiological) plasticity may be reasonable for the species studied, but the two cannot be regarded as mutually exclusive. In *D. melanogaster* there is evidence for a behavioural mechanism, since in extremely hot dry environments this species is difficult to trap in the middle of the day, and tends to be easier to collect in the early morning and evening. In the middle of the day, when temperatures may exceed 40°C in parts of Victoria, flies might tend to avoid this temperature by seeking cooler niches, emerging when the environment becomes more favourable. Levins (1969) found that *D. melanogaster* taken at midday at exposed traps in Puerto Rico were 5 per cent larger than those collected in the morning or late afternoon, and in protected sites there was no difference. In other words, only those flies which are probably genetically most resistant to desiccation appear at midday, by virtue of their large size (Parsons 1970a), and those that are genetically less resistant do not. Thus *D. melanogaster* shows genetic heterogeneity, physiological plasticity, and behavioural mechanisms in coping with hot dry environments, and it might be expected that in the various species all three may play a role, but that between species the relative importance of each may vary in different ways.

Turning to cold tolerance, laboratory work has indicated that varying degrees of tolerance to cold stress occur at different stages of the life cycle of *D. pseudoobscura*. Crumpacker and Marinkovic (1967) found, in a region with a cold winter and considerable diurnal fluctuation, that adults were more resistant to cold stress than were larvae or pupae, which in turn were more tolerant than eggs. It was also noted that adults collected immediately after the over-wintering period tended to be old individuals of the previous autumn. In *D. flavopilosa* (Brncic 1968) and *D. robusta* (Carson 1958a), the adults tend to hibernate, and it has been observed that the females of *D. robusta* caught in September have extremely small ovaries and very large quantities of body fat. It is inferred that these changes represent a biological adjustment of the species to over-wintering as adult flies (Carson 1958a). In *D. persimilis* (Mohn and Spiess 1963) both the pupal and adult stages may over-winter best, and in some other species it is likely that over-wintering may occur in non-adult stages (see Crumpacker and Marinkovic 1967). For example, Ives (1970) found evidence that larvae of *D. melanogaster* over-wintered in a continuously available rotten-apple pile in South Amherst, Massachusetts, a region where temperatures fall well below freezing in winter. These larvae become the progenitors of many of the flies emerging from the pile in June and July. In *D. subobscura*, immature stages were found to over-winter best, especially young larvae (Risch 1971). In all these cases habitat selection is likely to be of considerable importance especially in the more marginal environments. In *D. pseudoobscura*, for example, no super-cooling system which requires glycerol in the metabolic system of the insect (Salt 1961), and which would allow some release from a dependence on environmental sources, has been found. In *D. nigrospiracula*, however, which has as its host plant the saguaro (*Cereus giganteus*), super-cooling occurs such that the LD_{50} for one hour of cold treatment was $-7\cdot73\,°C$, the LD_{50} being the lethal temperature dose for a 50 per cent death-rate. The LD_0 was between $-5\,°C$ and $-6\,°C$, and the LD_{100} between $-9\,°C$ and $-10\,°C$ (Lowe, Heed, and Halpern 1967). In nature the flies have been found in a pre-dawn minimum air temperature of $-5\cdot1\,°C$, and began flying shortly after the sunlight reached them at the cactus. This enables the species to be more or less continuously breeding during the winter in a relatively cold-winter part of the Sonoran Desert.

Marinkovic, Crumpacker, and Salceda (1969) found substantial heterozygote advantage for cold-temperature resistance in *D. pseudoobscura*. For chromosome III the survival of homozygotes was 50 per cent less than that of the heterozygotes, and a much smaller differential was uncovered for chromosome II, although it was in the same direction. The pronounced heterozygote advantage associated with genes on chromosome III is another example of heterosis at sub-optimal environments (high or low temperatures and desiccation in ectotherms). An early observation of Heuts (1948) is in agreement in showing that longevity of adults in the presence of food at low temperatures ($0°-4\,°C$) is greater for *ST/CH* heterokaryotypes than the corresponding homokaryotypes (see discussion in §13.1).

2. Morphological traits in different environments

Several studies on morphological traits in different environments have been carried out. Thus, Misra and Reeve (1964) studied *D. subobscura* in populations extending from Scotland to Israel. The localities from which the populations came varied over 25° of latitude and just over 8°C in mean annual temperature. The populations had been maintained in the laboratory for between eight and eleven generations before being tested. Five dimensions, wing length and width, thorax length, head width, and tibia length showed strong positive correlations with latitude and slightly lower negative correlations with temperature (Table 8.2), therefore giving a very uniform cline in

TABLE 8.2

Correlations with latitude and temperature for females in D. subobscura *and* D. robusta

	Latitude		Temperature	
	D. subobscura	*D. robusta*†	*D. subobscura*	*D. robusta*†
Thorax length	0·729	−0·370	−0·546	0·315
Head width	0·604	−0·638	−0·500	0·646
Wing length	0·908	0·544	−0·685	−0·539
Wing width	0·796	0·188	−0·593	−0·205
Leg length‡	0·856	0·369	−0·712	−0·333
5 per cent significance levels	±0·576	±0·293	±0·576	±0·293

† The data for *D. robusta* came from Stalker and Carson (1948) and Stalker quoted by Misra and Reeve.
‡ Tibia length for *D. subobscura* and femur length for *D. robusta*.

After Misra and Reeve 1964

the five dimensions each increasing from south to north (see Prevosti 1955 for earlier data). This pattern differs from that found for essentially the same dimensions of *D. robusta* by Stalker and Carson (1948), where correlations with latitude were lower, and those for head and thorax size were of opposite sign in the two species (Table 8.2). A partial correlation analysis, and reference to selection experiments carried out by Robertson (1962), suggested to Misra and Reeve that there are two distinct groups of genes involved in these clines: first, a group of genes causing an increase in relative wing and leg size and responsible for the positive correlation of those dimensions with latitude in both species, and second a group of general-size genes causing correlated changes in all dimensions, which have been selected in opposite directions in the two species, with the result that head and thorax size increase in *subobscura* and decrease in *robusta* as latitude increases. In *D. robusta*, seasonal shifts in morphology were found such that there was a regular highly significant shift towards a southern morphology during the summer months from June to August (Stalker and Carson 1948, 1949). This shift is in the expected direction, assuming the southern morphology to have a high-temperature adaptive value. In *D. melanogaster* and *D. simulans*, Tantawy and Mallah (1961) found that wing and thorax lengths were greater at lower temperatures. It is therefore reasonable that there should be a seasonal change in morphological differences—for example, wing length declines during the summer months in wild Egyptian populations (Tantawy 1964).

In *D. melanogaster* and *D. simulans*, the number of flies with more than four scutellar chaetae was highest in Australia in a collection site with a Mediterranean climate, and lowest in a site with a subtropical environment (Fraser 1963). Since the flies on which the observations were made were cultured under standard conditions in the laboratory, it is reasonable to suggest that the variations found were not fortuitous but have some adaptive value. Perhaps the changes may be correlated with fly size, since it is known that, when flies of a given strain are grown under high temperatures, fly size and sternopleural and scutellar chaeta numbers are lower than at low temperatures (Parsons 1961, Fraser, Erway, and Brenton 1968). Therefore in natural populations, variations in fly size which can be induced experimentally in the laboratory are fixed in different regions presumably because of somewhat differing genetic architectures.

Levins (1969), using flies collected in the wild, found a body-weight cline for *D. melanogaster* in Puerto Rico such that flies caught at cooler, higher localities were larger than those taken in the hot lowlands. However, when raised under identical conditions in the laboratory, the coastal flies were larger—that is, the general expectation that phenotypic differences in nature are in the same direction as in the laboratory was contradicted in this case (but see later in this section).

Sokoloff (1966) studied *D. pseudoobscura* captured in nature for body-weight and/or wing length, wing width, and tibia length; he found that, within a given year, despite the wide range of ecological conditions prevalent in their areas of capture, flies were fairly uniform. There was, however, a distinct difference in size of flies captured in two successive years, and this has been correlated with climatic conditions prevailing in the two years. The uniformity within years allows it to be concluded that this species has adapted to a wide variety of environments by the evolution, in individuals belonging to the various races of this species, of favourable genotypes through natural selection which enable them to develop an optimal phenotype in the face of the wide range of environments. It can also be argued that clines might normally be expected as described for other species in the face of climatic variations, and it seems that different species adapt in different ways— especially as in *D. pseudoobscura* there are known and definite clines for chromosomal variants (see Chapter 12). In *D. pseudoobscura*, therefore, it seems that morphological variation may be at least partly independent of the types of inversions which characterize the various races of this species, since the morphological phenotypes do not seem to vary in the same way as the inversions, which are relatively environment-dependent in their frequencies. Even so, Anderson (1968) has reported that natural populations from Canada to Mexico vary in body size, such that populations from the Pacific coast are generally smaller than those from the interior provinces.

A problem in this work is that some of the observations are based on flies collected in the wild, and others are based on cultures set up from flies collected in the wild, often reared under relatively known and standard conditions. For most of the year, flies collected in the wild are usually intermediate between flies reared in the laboratory under conditions of maximal and minimal optimality, although during the summer the size of wild adults is lower, and this can be attributed to high temperatures (Tantawy and Mallah, 1961). This may be a partial explanation of Levins's (1969) data above, since under conditions of less stress in the laboratory coastal flies may be expected to be larger (but see also Levins, 1969 for further discussion).

Work on morphological traits *per se* thus allows certain inferences to be drawn about populations in relation to their environments, but it is difficult to be as precise as is the case for the work on the differential mortalities of genotypes caused by environmental stresses. Little work has been done in attempting to assess the selective significance of variations of morphological traits. Barnes (1968) set up two populations of *D. melanogaster* from the F_1 crosses of two inbred strains, Oregon and Samarkand, and maintained them at 25°C and 18°C respectively. Population sizes fluctuated around a value of 2000. The populations went to a sternopleural chaeta number of 23·2 at 25°C and 21·4 at 18°C after 600 days (it should be noted that these means indicate the opposite of what is expected to happen when flies are exposed to these temperatures in one generation). In each environment, maximum fitness as assessed by mean yield of progeny was associated with phenotypes around the population mean, and changes away from these values were accompanied by a reduction of fitness. A later experiment (Kearsey and Barnes 1970) based on genotypes extracted from a single wild population showed that in population cages, the variance of sternopleural chaeta number was much lower than when flies were raised under conditions of low densities. Therefore sternopleural chaeta number is subjected to stabilizing selection for the mean chaeta number in its given environment—although, of course, chaeta number can be varied environmentally; for example, a reduction of sternopleural chaeta number occurs under conditions in the laboratory of high temperatures or high competition (Parsons 1961). Unlike certain other of the traits discussed, a table given in Kearsey and Barnes (1970) does not indicate any real association with location of collection.

In conclusion, for some species and traits there is some evidence for an association between the size of morphological features and environment for flies collected in the wild. In other cases this is not so, and different species may adapt in different ways. Superimposed on this is the observation that when flies are grown in the laboratory under different environmental regimes variations occur. Finally, the form of natural selection to which these traits are subjected is likely to be stabilizing, but further investigations on this point would be of interest.

3. Insecticides and related compounds

The wide occurrence in nature of resistant insects after several generations of insecticide application is an example of rapid evolution promoted by the application of chemicals by man, and is also an economic problem. While before 1940 only eight insect species were known to have developed resistance, the number of resistant strains began a sharp upward trend after the introduction and use of DDT and other synthetic organic insecticides (Georghiou 1965). Thus, DDT-resistance has appeared in many insects and has been studied fairly extensively. The general conclusion is that insecticides form powerful selective agents, concentrating resistant mutants which were initially present in low frequencies in the original population. There is a substantial literature on the genetics of insecticide resistance (see, for example, Crow 1957, 1960, Oppenoorth 1965, Georghiou 1965) in many species, but here discussion is mainly restricted to *Drosophila*, especially *D. melanogaster*.

Some strains of house-flies are characterized by extremely high levels of resistance to DDT, but, in spite of the uncertainties in comparing levels of resistance in different species measured by different methods, Crow in 1957 considered that *Drosophila* had not then developed the high level of resistance of house-flies: this, presumably, is still generally true. In *D. melanogaster*, Tsukamoto and Ogaki (1954) and Tsukamoto (1955) studied larval resistance to DDT and other insecticides and found sizeable differences between various laboratory and wild strains. They grew mixed cultures in a medium containing an appropriate concentration of DDT, and counted the proportions of different types emerging as adults. Detailed genetic analysis revealed that a single dominant gene at about 66 on chromosome II was responsible for the resistance. Tests showed the same region to be resistant for BHC (benzene hexachloride), parathion, and PU (phenylurea), but sensitive to PTU (phenylthiourea). In addition, the resistant Hikone wild-strain larvae were resistant to nicotine sulphate. However, most of the nicotine-sulphate-resistance was due to a gene at 49 to 50 on chromosome III, although other parts of this and the second chromosome seem to be involved. Furthermore, this third chromosome locus was associated with PTU- and PU-resistance (see Ogita 1958, 1961). It is of interest that Georghiou (1965) reports that the only example so far of cross-resistance between a chlorinated hydrocarbon (DDT) and an organophosphate (parathion) is in *D. melanogaster*, and it has not been found in insects such as house-flies and mosquitoes. In *D. virilis*, dominant genes responsible for DDT-resistance are on the second and fifth chromosomes. The main effects of the two genes on resistance were almost equivalent (Georghiou 1965).

Crow (1954) found adult resistance to DDT to be polygenic in *D. melano-*

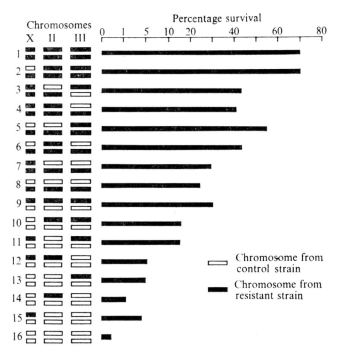

FIG. 8.3. Inheritance of DDT-resistance in adult *D. melanogaster*, showing polygenic control. (From Crow 1957.)

gaster. The procedure used was to grow a large mixed laboratory population in a cage whose inside surfaces were painted with irregularly increasing amounts of DDT, to develop a resistant strain. The cage was painted irregularly so as to try to simulate nature. Adults were tested by exposure for 18 or 24 hours to DDT residue on filter paper. Chromosome analyses were done using a sensitive control strain with genetically marked chromosomes. Each of the sixteen possible combinations, as females, was isolated between resistants and controls. From Fig. 8.3 it can be seen that the proportion of survivors increases with the number of chromosomes from the resistant strain. Statistical analyses showed that each of the major resistant chromosomes contributed to resistance. More detailed analyses revealed that there are probably several genes on each chromosome involved—that is, resistance is polygenic. Oshima (1954) obtained results similar in principle to those just described, and other studies cited by Crow (1957) are consistent with a polygenic interpretation, as is the case for adult resistance to BHC.

King and Sømme (1958) selected for DDT-resistance in *D. melanogaster* in population cages over a number of generations, such that after fifty or more generations of selection two populations derived from similar sub-samples of a single strain had an LD_{50} (see §8.1 for definition) of about twenty times the control. Genetic analysis again revealed genetic activity on the three major chromosomes. Crosses between the two resistant lines after twenty generations of selection gave an F_1 population with the same resistance as the parental lines, and an F_2 with significantly lower resistance with a mortality distribution of greater variance. It therefore seems that the two lines have achieved resistance by consolidating different combinations of genes for resistance, which again argues for a complex polygenic system involved in DDT-resistance (see also Merrell and Underhill 1956). Merrell (1965) reported on populations of *D. melanogaster* in which exposure to DDT had been gradually increased over ten years by 100–300 times the unexposed control; thus, the changes induced by DDT can be quite large.

In spite of the number of years since the first DDT-resistant strains were isolated, not a great deal more has emerged about the genetic basis of resistance, perhaps because of the difficulty of studying insecticide resistance in an active *Drosophila* laboratory, even though Oppenoorth (1965) commented that the distinction, characterization, and localization of single factors for resistance constitute a field for study. Such a development has been retarded in many species of insects by a lack of knowledge of their formal genetics, but this does not apply to some *Drosophila* species. The other field for study concerns the actual mechanisms involved. In many cases detoxication mechanisms seem to be involved. In fact three kinds of DDT-resistance with different genetic and physiological backgrounds are known in the house-fly (Grigolo and Oppenoorth 1966). These are: first, DDT detoxication by DDT-dehydrochlorinase (DDT-ase) dependent on a gene D-ase on chromosome 5; second, a different unknown detoxication resistance factor on chromosome 3, and third, knock-down resistance brought about by gene *kdr* on chromosome 2 of an unknown physiological nature.

The need to discover the actual mechanisms of insecticide resistance is emphasized by observations such as that of Ogita (1958) who found that strains of *D. melanogaster* as larvae resistant to DDT and other insecticides are susceptible to PTU. Ogita (1958) reported differences in percentage emergences of larvae grown on PTU medium, mainly controlled by two genes on chromosomes 2 and 3 respectively (see above), with a contribution from an incompletely dominant gene on the X. Various mutants are known to be sensitive to PTU, in particular ebony *e* (Kroman and Parsons 1960), and it is suggested that this may be because ebony larvae have less dopa-oxidase than normals, since PTU acts as a dopa-oxidase inhibitor (Lerner and Fitzpatrick 1950). Ebony homozygotes do not normally survive on 0·04 per cent PTU, whereas wild-type larvae survive on up to 0·12 per cent PTU (Parsons and Kroman 1960). Subsequently, Parsons (1963*a*) found ebony alleles segregating in various populations which had little or no effect on body colour, but which were as sensitive to PTU as were *ee* homozygotes. The polymorphism was found in various laboratory stocks, Oregon-R, Oregon-K, Kaduna (Nigeria), and Bikini, and in flies collected in the wild from Oregon and Cambridge, England. More recently, a search was made in seventeen Australian strains collected near Eltham, Victoria and set up from single inseminated females (Deery and Parsons 1972*a*). The seventeen strains were tested on PTU and, as for various traits discussed in §8.1, they were found to be genetically heterogeneous for response to PTU. From the three most sensitive strains, ebony alleles with no detectable effect on body colour were isolated, as were wild-type alleles from the three most resistant strains. Therefore, it seems likely that major differences within populations in PTU sensitivity are associated with alleles at the ebony locus. Perhaps there is some hope of attacking this problem physiologically and biochemically, because of the known effects of PTU as a dopa-oxidase inhibitor.

It is likely that a vast number of chemicals affect different genotypes differentially, so leading to population changes. Insecticides have been of prominence in this type of research because of their economic importance. Twenty years ago Herskowitz (1951) gave a comprehensive list, based on 314 published papers, of chemicals which had been reported to have genotypic or phenotypic effects on *Drosophila*, some of which may be of importance in the environment today. Among other examples, Robertson (1966) studied the adaptation of *D. melanogaster* to an initially unfavourable diet, the chelating agent ethylenediaminetetra-acetic acid (EDTA) in the food medium. The initial effect was to reduce survival and body size, and to increase development time, depending on the concentration. Populations grown on food containing EDTA adapted successfully, and some were able to grow on EDTA concentrations lethal for the original population. Genetic differences between one of the EDTA-adapted populations and the original population were studied by using marked inversions. Adaptation was shown to involve all major chromosomes, and there were substantial complementary interactions between non-homologous pairs of chromosomes. The adapation to EDTA was shown to involve high survival, faster development, and growth to a body size which is near the maximum attainable under optimal conditions.

4. Anaesthetics

Although of no known ecological effect, the anaesthetic ether has been studied in some detail (Deery and Parsons 1972b), from the point of view of genetic variability in *D. melanogaster*. However, studies using such chemicals may be of significance for the study of chemicals of ecological significance, such as insecticides, as they may provide information on the likely genetic

TABLE 8.3

Mean percentage mortalities of thirty flies 24 hours after etherization in a 4×4 diallel cross between four strains from Leslie Manor, Victoria

		Mortality in females				Mortality in males			
		Strain of ♂ parent							
		32	25	30	26	32	25	30	26
Strain of ♀ parent	32	7·74	12·90	11·29	20·81	28·79	43·52	49·84	61·71
	25	10·89	27·95	16·15	23·33	29·24	59·63	28·59	37·10
	30	10·00	14·79	65·63	47·58	54·37	69·64	78·34	84·28
	26	11·29	23·01	76·37	70·49	58·07	54·20	76·67	79·31

		Analysis of variance (F values)	
		Females	Males
Source of variation	d.f.		
General combining ability	3	50·80†	13·89†
Specific combining ability	6	9·68†	2·05
Reciprocal effect	6	2·12	2·33
Error	16		

† $P < 0.001$.
Etherization was carried out for two minutes under standardized techniques.

After Deery and Parsons 1972b

architectures involved. In *D. melanogaster* Rasmuson (1955) found evidence for a cytoplasmic effect for resistance to ether, while Ogaki, Nakashima-Tanaka, and Murakami (1967) found no such evidence, but an ether-resistant strain was found to have major genetic activity at 61 on the third chromosome, and minor activity on the X and fourth chromosomes. Deery and Parsons (1972b) studied the ether tolerance of adult flies 24 hours old in two Victorian populations, Leslie Manor and Eltham, based on fifteen strains set up from single inseminated females in the former locality and seventeen in the latter. As for the various traits cited in §8.1, genetic heterogeneity was found between strains indicating genes segregating in the wild for variations in ether sensitivity.

Diallel crosses were then set up between extreme strains; an example for four Leslie Manor strains is presented in Table 8.3. The results were analysed according to the model of Griffing (1956) (see §5.3). In both sexes there are large and significant general combining abilities indicating additive differences between strains. Specific combining abilities were much smaller but were significant for females. It is clear that for any crosses in which either of the ether-resistant strains are involved, i.e., 32 or 25, percentage mortalities were considerably lower than in those crosses between purely sensitive strains. This indicates that the ether-resistant strains show some dominance over the sensitives, which is in partial agreement with Ogaki *et al.* (1967) who found resistance to be completely dominant. Similar results were obtained for a 4×4 diallel cross between four Eltham strains.

The genetic analysis was taken further, first by locating genetic activity to chromosomes using the technique of Kearsey and Kojima (1967; see also §6.2), and then to regions of chromosomes using recessive marker stocks. Based on two extreme strains, one resistant and one sensitive, major genetic activity was found on chromosome III and to a lesser extent chromosome II. Relative to the marker stock, on chromosome II there was genetic activity for sensitivity at the proximal end. On chromosome III there was a region of weak resistance on the proximal arm and strong sensitivity on the distal arm. Therefore, based on these two strains alone, there are indications of genes controlling variations in resistance and sensitivity in various parts of chromosomes II and III. If other strains were taken, no doubt different regions of genetic activity would emerge. The genetic architecture of ether resistance (and sensitivity) can thus be assumed to be quite complex, and it would be of interest to study it further.

For chloroform resistance, as for ether resistance, differences were found between strains. A diallel cross with four extreme strains revealed mainly additive genetic activity. However, there was little or no correlation between strains for sensitivities to the two anaesthetics. This is not unreasonable, since the two anaesthetics differ chemically, ether being C_2H_5—O—C_2H_5, while chloroform contains Cl atoms, its formula being $CHCl_3$. Furthermore, the mortality curves obtained after exposure to the two anaesthetics differ, since etherized flies die rapidly, usually under the anaesthetic, there being little difference in the number dead at one hour or 24 hours after etherization. For chloroform, death occurs progressively over the 24-hour period so that, for the doses used, almost all flies were dead after 24 hours, but at four hours after exposure to chloroform, the above-described differences were found between strains.

5. Genetic architectures under chemical stresses

Ether and chloroform resistance were found to be mainly controlled by additive genes from the diallel crosses, and similar evidence was found for PTU. Fig. 8.3 (see page 101) shows a reasonable level of additivity for DDT-resistance as the number of resistant chromosomes is increased. Additive genetic control is convenient from the point of view of detailed genetic analyses. There have been several papers arguing for a correspondence between the genetic architecture of a quantitative trait and the type of selection to which it has been exposed in the past (Breese and Mather 1960, Mather 1966). Thus, directional selection will lead to a quantitative trait showing directional dominance and a duplicate-type gene interaction, as found for traits selected for uniformly high fitness such as viability (Breese and Mather 1960) and hatchability (Kearsey and Kojima 1967). Stabilizing selection, however, leads to little dominance and weak interactions, and such dominance and interactions as do occur tend to be ambidirectional. The genetic architecture of ether resistance probably corresponds more to this latter situation. It is at present difficult to say more, especially as ether is presumably not a stress normally present in wild populations. These comments probably apply to many other chemical stresses not normally present in the environment.

The approach just described to studying chemical stress traits seems useful in providing a first over-all view of the genetic architecture of a trait, from which more detailed studies can proceed. Such studies may consist in carrying out directional selection on the extreme strains, or on hybrids between extreme strains (see Hosgood and Parsons 1967*b*, Lee and Parsons 1968). For traits controlled by additive genes, responses to selection are likely to be rapid. Following this, loci controlling the trait under study could be located and studied from various points of view—for example, physiologically or genetically. The detailed study of how various chemicals act would be considerably assisted by locating as many genes as possible controlling resistance to the chemicals and forming extreme homozygous stocks.

A genetic architecture of several loci of reasonably large effect, which are mainly additive—as seems likely for many chemical stresses—has implications for insect control. Traits normally under stabilizing selection would be expected to respond to directional selection rapidly by the rearrangement of these additive genes. The application of a chemical stress is a form of directional selection of the type used by animal and plant breeders to change a trait in a desired direction. The actual response will depend on some of the factors discussed in Chapter 6, but especially on the number and distribution of the genes (i.e., linked or unlinked), and their gene frequencies in the base population. This means that in many cases a rapid build-up of resistance to insecticides may occur. Traits with a direct and obvious relation to fitness, such as viability and hatchability, are under continuous directional selection to increase fitness. Therefore under such circumstances responses to directional selection would be limited. To be effective, a new insecticide must be a chemical to which an organism has not, or has hardly, been previously exposed, which means that strong directional selection for resistance is scarcely likely to have yet occurred. As soon as the insecticide is applied, rapid responses to selection are likely provided that a few resistants survive. From the few resistants there will be a rapid build-up of resistant strains, which implies progressively higher doses for control in subsequent generations. This can proceed to the level where the insecticide becomes ineffective. The alternative is to change to another insecticide, for which, in the same way, there would be a rapid build-up of resistant strains. So long as the chemist continues to produce enough new chemicals, the insects may perhaps be kept under control, but this does point to a continuing contest between insects and man, with rapid evolution occurring in the insects as a result of it. In the meantime, the whole environment is being polluted, since substances such as DDT have been found in the tissues of many animals as diverse as fish, mice, and men. The increase in DDT-tolerance by a factor in excess of 100 in *D. melanogaster* shows the type of difficulty involved.

Before changing to another insecticide, it is desirable to seek evidence on possible cross-reactions, as discussed earlier in this chapter. To test for this possibility, highly resistant strains for one insecticide can be tested for the other, or alternatively the approach of testing strains set up from single inseminated females in a wild population for the two insecticides can be used. This approach was used by Parsons (1970*b*) for possible associations between sternopleural and abdominal chaetae. Over-all, the data showed a weak correlation between the two traits. However, considering this by strain, the correlations range from negative (but not significantly <0) to positive (but significantly >0). It was concluded that, just as the response to directional selection was extremely rapid when based on strains extreme in the direction of the required selection response, a similar process may be useful for directional selection based on two traits simultaneously, by using only those strains showing a positive correlation. The likelihood that selection for one insecticide might lead to a correlated response to a second can presumably be assessed by tests of this nature, but more work is needed on this point.

6. Variations in chemicals normally present

So far, chemicals implying a definite stress but which are not normally present have been considered: there is also evidence for variations in the tolerance to chemicals which regularly occur in the nutritional environment of *Drosophila*, but in abnormal quantities. For example, sodium salts such as NaCl are present in the normal *D. melanogaster* diet. However, in the presence of 1·0 M NaCl in the culture medium, considerable differences in emergence rates were found between both mutant and wild-type strains (Miyoshi 1961). Excesses of NaCl cause inhibition of growth and eventual death. By continuous selection, tolerance to NaCl was increased. In one strain the increase was from an emergence value of 50·2 per cent on 1·0 M NaCl to 90·6 per cent at the ninth generation of selection, and this level was then maintained, even on culture in media supposedly free of NaCl. Waddington (1959) developed highly resistant strains by selection, and reported that the anal papillae of larvae resistant to high concentrations of NaCl were larger in size than those of the susceptible ones. The function of this organ is not exactly known but it is supposed to be concerned with osmotic regulation. Miyoshi (1961) did not find any morphological changes.

Sang (1956) carried out a thorough analysis of the quantitative nutritional requirements of *D. melanogaster*, and has developed a suitable synthetic medium which allows larval development in a wild-type strain to be completed in about 4·4 days at 25°C, as compared with 4·1 days when the strain is reared on an optimal supply of killed yeast. Deficiencies of essential nutrients do not produce particular 'disease' syndromes, but may lead to death during the larval stage, or during the developmental crisis of the pupal instar. Whatever adults emerge, their appearance is normal although with reduced size, but melanotic tumours are often found. Sang considers that the adaptability of larvae to survive when confronted with dietary deficiencies is an aspect of the ability of larvae to survive under intense overcrowding which may frequently occur in nature. The development of a specific synthetic medium has facilitated the study of variables such as body-weight and developmental time on different diets, and also of the effects of selection for body size on different diets. Strains grown on media deficient in protein and nutrients in general show a reduction in body size; even so, body size can be increased again by selection (Robertson 1960*a*, *b*). Church and Robertson (1966) developed lines differing greatly in body size and/or development time by selecting for either large or small body size, or shorter or longer development time on different chemically defined axenic media. The ten lines reported in this particular paper were selected either on an optimal medium or on media deficient in either protein, ribonucleic acid, or choline. Comparisons for the contents of various constituents were made at a critical size, namely the time in early third instar when larvae can complete development even if no longer allowed to feed, and in newly hatched adults. There was a high positive correlation between DNA content and the time taken to reach the critical size and also the time to pupation. Protein and RNA content varied within wide limits without alteration of development time. The association between DNA content and development time which underlies the correlation between body size and development time was found to be related to the rate of DNA synthesis in early life, such that the rate of synthesis tends to be negatively correlated with the absolute DNA content at the critical size and in the adult. The biochemical analysis of growth in relation to nutrition is clearly of ecological significance, but as yet it seems difficult to relate the work to natural populations. However, Robertson (1963) considers that variations in larval growth rates according to nutrition provide a flexible system for adjustment to different ecological conditions.

9
COMPETITION

1. Introduction

Two or more organisms coexisting in proximity may influence each other as they may share a resource that is in short supply. Such an influence is variously referred to as competition or interference, etc., but here we shall use the term competition (see Sammeta and Levins 1970). Competition implies a similarity in the requirements of the competing individuals. Thus Birch (1957) has written 'Competition occurs when a number of animals (of the same or of different species) utilize common resources the supply of which is short; or if the resources are not in short supply, competition occurs when the animals seeking that resource nevertheless harm one another in the process.' As the genetic resemblance between two individuals increases, so, too, will the similarity of their requirements, which in turn will be expected to lead to more intensive competition between those two individuals. It is reasonable, therefore, that the most intense and widespread competition would be expected to be within species and varieties, and that between-species competition would be of lesser importance. However, between-species competition would be expected to be most intense between those that are most closely related (Sokoloff 1955). As is shown in Chapter 14, the understanding of competitive relationships between closely related species and its appreciation is of considerable evolutionary importance. This chapter considers some of the principles emerging from competition experiments between genotypes *within* species. Further examples appear later, especially in Chapter 12, for species of *Drosophila* that are regularly chromosomally polymorphic.

2. Genotype and environment in competitive ability

Not many experiments have been carried out on the effects of genotype and environment on competitive ability in *Drosophila*, and indeed much of the definitive work has been done in plants. Thus Harlan and Martini (1938) grew eleven barley varieties in a mixture, and studied the proportions of these varieties at various localities over a number of generations. In all localities, one variety (which differed between localities) quickly became dominant while others declined to very low proportions. These and other data suggested that there are marked differences in competitive ability between varieties. Sakai (1961) provided direct proof of the genetic basis of competitive ability by comparing the performance in pure culture and in mixtures of a number of cereal varieties. Nine wheat varieties were grown alone and in all possible pairs to form a diallel cross, from which general and specific combining abilities were estimated. The varieties showed highly significant differences for both. Sakai and his associates also found differences in intraspecific competitive ability for varieties of barley and rice. Studies on the inheritance of competitive ability in rice revealed low heritabilities for various measures of competition (Oka 1960, Akihama 1968). This may be expected since in cultivation a plant species would have been exposed to strong directional selection for most measures of competitive ability, so leaving rather little additive genetic variation. However, in new environments, or in progeny derived from crosses, the situation may well be different (see Mather 1961).

Therefore in *Drosophila*, it is likely that heritabilities would be low and such genetic control as might be found would not be additive to the extent of the chemical stresses discussed in §8.5. This is because natural selection is likely to be directional, favouring higher competitive ability, rather than stabilizing. Mather and Cooke (1962) set up experiments designed to test for differences in competitive ability in *D. melanogaster* among the nine genotypes derivable from four lines representing all the true-breeding combinations of chromosome III, with chromosomes X and II treated as a unit, from an inbred Samarkand (S) strain, and a Birmingham (B) strain of an Oregon inbred strain. The four lines can be written

$$(B/B \ B/B) \ B/B \quad \text{line 1}$$
$$(B/B \ B/B) \ S/S \quad \text{line 2}$$
$$(S/S \ S/S) \ B/B \quad \text{line 3}$$
$$(S/S \ S/S) \ S/S \quad \text{line 4,}$$

where the brackets enclose chromosomes X and II which were treated as a single unit, and chromosome III is outside the bracket. The five possible different F_1's from these lines with the above four lines provided the nine genotypes. Crowded mixed cultures were raised at two temperatures from all possible combinations of the nine genotypes taken two at a time, and the proportions of offspring of the two genotypes were estimated from the average number of sternopleural chaetae in both sexes, since the various genotypes were characterized by different chaeta numbers. Measurements were obtained of the relative competitive abilities of both genotypes at 25 °C and 18 °C, and significant differences were observed. Analysis of the effect of chromosome III, and of the chromosome X-II unit, indicated genic interactions of a duplicate type, which is to be expected for a trait under directional selection. Some differences between temperatures were also found, since chromosomes X and II appeared to have a similar joint effect only at 18 °C.

Gale (1964) studied three strains which differed in competitive ability, namely *vg* (vestigial), *dp* (dumpy), and *Or* (Oregon). The former two were selection lines and were assumed to be reasonably inbred, and the latter was an inbred strain. All three gave consistent performances for competitive ability. It was found that *vg* was superior to *dp*, and *dp* to *Or*. This held for three food volumes used, namely, 4, 6, and 8 ml. Assuming competition is for food, from a knowledge of the yields in pure culture and in two-way mixed cultures, it is possible to predict the result actually obtained when all three strains are competing. This suggests that competition is mainly for food. Gale considers that the differences are probably fairly simple, probably representing feeding-rate differences for food. In fact, Bakker (1961) has shown that the competitive superiority of wild-type over Bar larvae is due to the fact that wild larvae feed more rapidly than Bar larvae, so obtaining a disproportionately large amount of the limited quantity of food available. Kearsey (1965*a*) made the assumption that larval competitive ability depends on feeding-rate by limiting a resource, killed bakers' yeast; the medium being an agar-yeast medium. Based on this assumption, the viabilities of two competing strains in mixed culture can be predicted from their viabilities in pure culture at any food level. For two strains, the predicted viabilities are such that the difference between the total yield of mixed and pure cultures should follow a cubic regression over the range of food levels limiting viability. An experiment in *D. melanogaster* designed to test this prediction was set up, and the expected cubic relationship was found, for a wild-type inbred strain and a *vg* strain selected for sternopleural chaeta number and then inbred for twenty generations.

The evidence presented so far shows that increasing the density of a population brings about competition among individuals and so lowers the proportion of zygotes reaching maturity. Mather (1961) suggested that at low densities individuals may in fact co-operate in maintaining a minimum needed to prevent the environment from becoming unsuitable to all, and it was thought that such an effect may be apparent in underpopulated cultures. Lewontin (1955) in *D. melanogaster*, and Lewontin and Matsuo (1963) in *D. bucksii*, have reported results consistent with the co-operation hypothesis. Lewontin and Matsuo studied a wild-type strain and the four mutants (white, claret, cut, and yellow) in isolation, and in mixtures for larval densities of 2, 8, 32, 128, 256, and 512 per constant amount of food. The viabilities in all cases gave curves that are concave downwards, in other words, both high and low densities resulted in lower survival, which argues for co-operation at low densities and increasing competition at high densities (Fig. 9.1). No live yeast was used in these particular experiments, hence undercrowding could not have been due to a lack of conditioning of the food for yeast growth by the action of the larvae: the co-operative action must lie elsewhere. Kearsey (1965*b*) took an inbred Oregon stock in which fungal contamination, in the form of a strain of *Penicillium* from old fly cultures, was deliberately introduced into the experimental vials. Eggs in densities of from 1 to 20 per vial were prepared for three replicated experiments. The curves showed low survival for low egg densities, which then increased with egg density, indicating co-operation. The survival then fell in two experiments at high egg densities and was probably near to this in the third, thus indicating competition at the higher densities. The over-all differences in average viability between the three experiments probably reflect both variations in the moisture content of the medium which would affect hatchability and the extent of fungal contamination. The results are therefore consistent with three processes occurring as population densities increase, namely, co-operation, neutrality, and competition.

Another variant of co-operation is the mutual facilitation that has been shown experimentally to occur between strains of *D. melanogaster* for larval

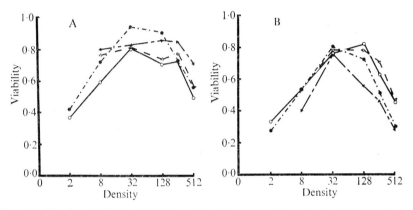

FIG. 9.1 Absolute viabilities of (A) Acme (a wild-type stock) and (B) white in *D. bucksii*.

Solid line	−100 per cent Acme	100 per cent white
Long and short dash	−75 per cent Acme	75 per cent white
Short dash	−50 per cent Acme	50 per cent white
Dot-dash	−25 per cent Acme	25 per cent white

(From Lewontin and Matsuo 1963.)

viability, perhaps attributable to the effect of diffusible metabolic products from one genotype on another (Dawood and Strickberger 1969). Mutual facilitation may be of importance in polymorphic situations where the different genotypes segregating in the population could have slightly different ecological requirements so that mutual facilitation could result. However, not all inter-genotype interactions increase survival. Weisbrot (1966) studied the interaction of wild-type and mutant individuals of *D. melanogaster*, and of wild-type *D. pseudoobscura*. The survival of larvae of different types was determined when these developed on a certain culture medium, and on the same medium 'conditioned' by the presence of killed larvae of the same or a different sort. A number of interactions were observed, some killed larvae increasing the survival of certain other larval genotypes, and some decreasing the survival of other genotypes, while in some cases there were no effects. It appeared as if inhibitory actions (presumably through the action of metabolic waste products) are about as common as facilitation. Perhaps, however, among naturally occurring coexisting genotypes, facilitation would be more common as a result of natural selection (for example, in polymorphic situations).

Mutual facilitation would be one reason why modification of fitnesses occurs between genotypes, the exact result being unpredictable on the basis of the genotypes tested in isolation (Lewontin 1955, Lewontin and Matsuo 1963), in contrast with some of the results cited so far. It is of interest that Lewontin and Matsuo (1963) found that the relative viabilities observed in mixtures agreed with predicted viabilities from pure cultures at intermediate (optimal) densities, but deviated at high and low densities. Thus at these extreme densities, the genotype with the higher viability in pure culture often had a lower viability in mixed culture. Especially at high densities, it was the proportion of the two genotypes in the mixtures that was important in determining viabilities.

Another manifestation of increased competition is the increased variability of traits. Robertson (1960a) showed this by growing larvae of *D. melanogaster* on progressively more deficient diets, and found the within-culture variance for body size to be twice as great on a low-protein diet as on the usual live-yeast medium. (Reduction of diet is, of course, associated with increased competition.) He attributed this increase in variance to the segregation of genetic differences which are not expressed under more favourable conditions, when they contribute little to the variance under these conditions. Parsons and Kroman (1960) reported the variability of emergence times in *D. melanogaster* to be generally higher for cultures set up from 100 newly hatched larvae as compared with 25. The increase in variance was particularly marked for flies grown on food containing either of the two chemicals PTU and $AgNO_3$, and Parsons (1961) showed increased variability of fly weight under PTU-stress. At 30°C, sternopleural chaeta number variability was increased as compared with 25°C but fly weight variability was not affected. Therefore genotypes under stress, whether because of competition, temperature, or chemicals, may often show more variability compared with more optimal conditions.

3. Competition and natural selection

Competition is one of the ways by which natural selection can take place. The two terms are not synonymous, since selection can take place in the absence of competition through other fitness components such as mating behaviour, fecundity, and adaptation. Competition presumably may act to increase or decrease these differences, or even to reverse them. In plant-breeding programmes, information on the nature and extent of the relationship between competitive ability and productivity (as grain yield) is critical. Sammeta and Levins (1970) reviewed this and reported that in some cases the correlation is positive and in others negative. Gale (1964) studied the fertility and progeny per fertile mating in his study of vg, dp, and Or, from ex-competitors at a reasonably intense level of competition (90 larvae in 8 ml. medium corresponding to 0·78 mg. yeast per larva as mixed cultures, there being 45 larvae per strain). Single pairs of ex-competitors were set up and allowed to produce their maximum number of progeny. Table 9·1 shows

TABLE 9.1

Single-pair crosses of ex-competitors

Line	Fertile	Sterile	Total matings	Mean progeny per fertile mating	Total progeny for given number of matings (40)
Or	3	17	20	55·0	330
dp	9	31	40	94·7	852
vg	19	21	40	62·2	1182

After Gale 1964

fertility to be in the sequence $vg>dp>Or$ which is in direct correspondence with competitive abilities. The mean number of progeny per fertile mating gives $dp>vg>Or$, but fitness as a whole over the total number of matings, fertile and infertile alike, should be considered. This gives again $vg>dp>Or$, so that in spite of differences in the *numbers* of progeny per fertile mating, the total progenies for a given number of matings are of the same order as the fertilities. Therefore, in this case there is an association between competitive ability and certain other fitness components, but rather few studies have been carried out, and it would certainly be unwise to assume *a priori* a positive association between fitness in general and competitive ability. For example, hatchability showed no association with strain.

Gale (1964) reported that a strong competitor suffers more in a homogenous state than in a mixture with weak competitors. Furthermore, replacement of strong competitors by weak competitors in the initial composition of a mixed culture led to increased viability of both competing strains. In a mixed population each strong competitor interferes with fewer strong competitors than when homogenous, thus we have a likely density-dependent situation (see §7.2 for density dependence involving mating) where the strong competitor derives an advantage at the expense of a weak competitor. Another suggestive example of a dependence of viabilities on the proportions of competing genotypes comes from Parsons (1959a), who made a similar observation in two factor linkage crosses according to whether they were carried out in the coupling or the repulsion phase. In *D. bucksii*, Lewontin and Matsuo (1963) compared pairs of genotypes in ratios of 4:0, 3:1, 2:2, 1:3, and 0:4 for larval competition experiments. For the highest larval densities, 256 larvae per vial and 512 especially, there was a real effect of proportions on the viability of both genotypes in mixtures of genotypes, showing again density-dependent effects. At lower densities there were some examples, but, as the authors point out, the analysis they used was not very powerful at such densities.

Some recent results of Huang, Singh, and Kojima (1971) are of considerable interest, especially if they can be generalized. These workers studied the viabilities of genotypes at the esterase-6 locus of *D. melanogaster* under the control of two alleles Est-6^F and Est-6^S, abbreviated as F and S respectively. Fifty first-instar larvae of each of the genotypes FF, FS, and SS were placed in vials and, when they had reached the pupal stage, were removed or killed. Following this, 150 first-instar larvae of each genotype were grown in the media conditioned by the above method until all adults had emerged. Nine viability coefficients were obtained from the ratios of emerged flies to input larvae (Table 9.2). Clearly, the lowest viabilities are for larvae grown in their

TABLE 9.2

Viability coefficients for various combinations of genotypes at the esterase-6 locus in D melanogaster *obtained by using conditioned media*

Genotypes	Conditioning genotypes		
	FF	FS	SS
FF	0.923 ± 0.053	1.068 ± 0.056	1.130 ± 0.058
FS	1.090 ± 0.057	1.000 ± 0.055	1.087 ± 0.057
SS	1.146 ± 0.059	1.078 ± 0.057	0.928 ± 0.053

After Huang, Singh, and Kojima 1971

own conditioner genotype. In experiments where the frequencies of the genotypes were varied, the relative viability of a homozygote increased as its frequency decreased in media conditioned by its own genotype. The conclusion is that individuals of a conditioning genotype either deplete nutrients or leave metabolic products harmful to their own genotype. The authors suggest that fitnesses are dependent on allele-frequency so that rare genotypes have higher fitnesses, and abundant ones lower fitnesses, in relation to the equilibrium frequency of the alleles; thus, the mode of selection is frequency-dependent. They argue that it may be common that near the point of equilibrium genotypes might be expected to tend towards neutrality for fitnesses (see §7.2 for similar evidence from mating experiments).

In nature, genotypes or species are found mixed together in various proportions and in various environments; they would therefore be susceptible to competitive stresses of various kinds according to the genetic types, their proportions, and the environment. In particular, the role of the proportions of genotypes seems to be assuming increasing importance over the last few years. The issue of environment has been explored less, but relevant discussion appears in Chapter 13. It is indeed difficult to avoid the conclusion that competition in the broad sense must enter into most of the forms of selection that impinge on populations in the wild, and that it must play its part in determining most genetical adjustments and equilibria. This differs from classical models of the effect of selection on gene frequencies in populations, where it is customary to assume that individuals and genotypes have no differential effects on one another's fitness (Mather 1969). Competition in its widest sense, which includes co-operation (positive or negative), and density-dependent effects, is clearly in need of much more study, especially as it is a phenomenon making many of the classical theoretical models of population genetics appear to be oversimplifications.

10
DISPERSION AND MIGRATION

1. Dispersion

THE movements or the resting point of an individual fly in the wild cannot be predicted in advance: it is, however, possible to describe the distribution of movements of flies and their ultimate destinations. Such information has been sought by releasing large numbers of marked flies and noting their positions when recaptured, thus enabling frequency distributions of dispersal distances after a lapse of time, at various distances from the place of origin, to be calculated. Dobzhansky and Wright (1943) have presented thorough and systematic analyses of the dispersion of *Drosophila*, and a simplified analysis is given by Wallace (1966a, 1968). Here Wallace's presentation, being simpler than the original, is given.

During the summers of 1941 and 1942, the dispersion of *D. pseudoobscura* was studied on Mount San Jacinto in Southern California. Each experiment consisted of the release of several thousand adult flies, at a given point on one day, which were marked with the eye-colour mutant orange. This was followed by the subsequent daily recapture and release of these marked flies at regularly spaced traps. In this way the day-to-day distribution pattern of the released flies could be reconstructed from the numbers of marked flies recaptured at various distances from the point of release on different days. Wallace's (1966a, 1968) re-analysis of the data showed that the logarithm of the number of marked flies recaptured at various distances from the point of release decreased linearly with the square root of the distance (Fig. 10.1). The regression coefficients for this relationship came to -0.1447 and -0.1160 for the first and second days after release, respectively. For the third and fourth days, they came to -0.1031 and -0.0940, in other words, the flies were tending to spread further from the release point. The linearity of the distribution in Fig. 10.1 makes it easy to compare with experimental data. It should be noted, however, that Wright (1968) has suggested that variations in the results of separate experiments may be responsible for the linearity. Dobzhansky and Wright's data show that as many as one-quarter of all flies in a single highly localized collection may have emerged from pupae within 25 yards of the collection site (Wallace 1966a).

The next question that can be asked is whether all species of *Drosophila* disperse equally rapidly. A number of studies have been made using several *Drosophila* species. The first of these was by Timofeef-Ressovsky (1940) and

Timofeeff-Ressovsky and Timofeeff-Ressovsky (1941a, b) with *D. funebris*. Perhaps because of the choice of release point (the centre of a village refuse heap), the actual dispersion of these flies was very slight. However, the pattern of dispersal was the same as that described for *D. pseudoobscura*, in that the logarithm of numbers of marked flies recaptured declined linearly with the square root of the distance from the point of release. For all the flies caught throughout the course of the recapture experiment, the regression coefficient defined as above came to -0.4563 for *D. funebris* and -0.1020 for *D. pseudoobscura*, showing the low dispersal rates of *D. funebris*. It can, in fact,

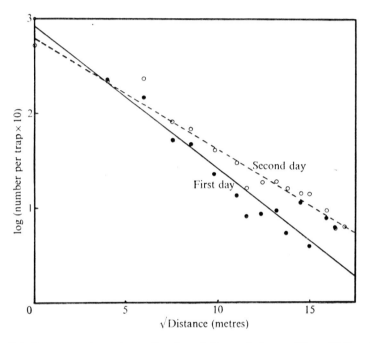

FIG. 10.1. Recapture of experimentally released *D. pseudoobscura*. (From Wallace 1968.)

be shown that the number of recaptured flies decreased by 90 per cent in 100 yards for *D. pseudoobscura*, but in five yards for *D. funebris*. Data for *D. willistoni* given by Burla, da Cunha, Cavalcanti, Dobzhansky, and Pavan (1950) gave a regression coefficient of -0.2306 and a distance of 15–20 yards for a 90 per cent decrease in the number of recaptured flies. Dobzhansky and Wright (1943) have described a small experiment using *D. melanogaster*, in which according to Wallace (1968) it seems that *D. melanogaster* disperses in a manner similar to *D. willistoni*.

The model used where the logarithm of the number of released flies decreases linearly with the square root of the distance from the point of release is expressed in terms of the proportion of flies in each collection that has arisen within a given distance of the collecting site in Fig. 10.2 (after Wallace 1966a), which shows the extremely small distances moved, especially in *D. funebris*.

Richardson (1969) studied migration in *D. aldrichi* with an array of baits of cantaloup (*Cucumis melo*); in the central bait a chemical label, dysprosium acetate, was used. Previous experiments showed that dysprosium acetate is an effective label and is non-toxic at the concentration employed. Baits, including the labelled one, remained in the field for seven days with no

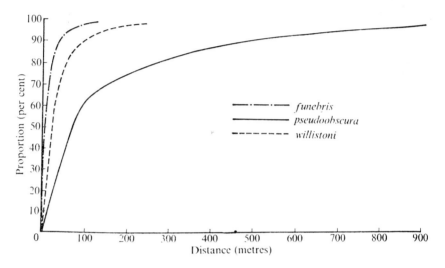

FIG. 10.2. Reconstruction of the composition of local collections of flies based on dispersal data. The curves represent the proportion of flies in each collection that has arisen within a given distance of the collecting site. (From Wallace 1966a.)

collecting activity, and on the eighth day flies were collected with a net in the early morning; this procedure continued for 14 days. The dispersal rates were low, no labelled flies out of thirty-two captured being collected at 36 metres. Richardson considers that this technique is preferable to that used in some of the earlier experiments already discussed, in that no flies are released into the natural population to complicate the interpretation by possibly increasing dispersal rates by initial crowding. Wallace's (1966a, 1968) analysis of Dobzhansky and Wright's (1943) data in fact shows a bias towards increased dispersal activity soon after the release of marked flies. Indeed, Wallace considers that the technique of release and recapture generally leads to overestimates of dispersal rates.

It is also of interest to note that other organisms show dispersal patterns compatible with the model discussed (Bateman 1950). In human populations, marriage patterns can be thought of as indicators of the dispersion of human beings, the primary data being obtained by assessing the birthplaces of the couples, and then calculating the probability of marriage as a function of distances between husband and wife. Cavalli-Sforza's (1962) analysis of Italian data gave a reasonable fit to the model, except that those individuals born very close together tended to marry more frequently than expected on the model.

It has been shown experimentally (Chapter 12) that a sizeable fraction of chromosomes of *Drosophila* populations carry recessive lethal genes. Random combinations of such lethal chromosomes are not usually lethal to the flies carrying them, since in such combinations the lethal genes do not usually occupy the same locus. However, on occasions, combinations of lethals of seemingly different origins do kill their carriers, and in this case the lethal genes presumably occupy the same locus. When lethal chromosomes are obtained from geographically restricted areas, the frequency of allelism is greater than that observed when the collection sites are more distant from one another. There is a certain probability of allelism due to the finite number of loci at which lethals can occur, and to this low constant frequency of allelism can be added another source of allelism, namely, identity by common descent. Therefore, in restricted localities, two chromosomes may carry identical lethals because each is a direct descendant by replication of a single ancestral lethal gene. It is not therefore surprising that the frequency of allelism is higher among lethal chromosomes from a single locality than among those collected at more remote distances. Dobzhansky and Wright (1941) and Wright, Dobzhansky, and Hovanitz (1942) showed clearly for flies collected on Mount San Jacinto in Southern California that the smaller the area from which the lethal chromosomes were obtained, the higher the probability of allelism (see Table 10.1). The table also shows that time can be substituted for distance, in that the more nearly simultaneously two lethals are taken from a restricted locality, the more likely they are to be allelic.

Now since lethals are carried from place to place by dispersing flies, it follows that the logarithm of the frequency of allelism due to common descent should decrease more or less linearly with the square root of distance as found for the actual dispersion of flies. Wallace (1966*b*) tested this suspected relationship in *D. melanogaster* males collected near Bogotá, Colombia for four trapping sites that were spaced linearly at 30-metre intervals. The second and third chromosomes were those tested. The lethals at the four sites were inter-crossed within and between collecting sites and gave an average frequency of lethals of 0·0461 within sites, and 0·0365, 0·0324, and 0·0275 for those sites 30, 60, and 90 metres apart respectively. In other words, the chance of allelism declined with increasing distance between sites, which indicates that the lethals spread from locality to locality as do the flies themselves, and the regression analysis of the logarithm of the frequency of allelism on the square root of distance gave a regression coefficient of -0.0262 ($P<0.02$ for deviation from 0). Thus, the analysis of the allelism of lethals is a technique by which the diffusion of genes through natural populations can be studied and, like Richardson's (1969) chemical labelling procedure, it has the advantage that flies are not introduced into the wild population, although of course the procedures for testing are necessarily tedious, involving the detection and testing of lethal chromosomes.

A further problem, not considered in much detail for flies introduced into natural populations, is the question of fitness differentials between the flies introduced and those in the population. Thus, the experiments of Dobzhansky and Wright (1947) in *D. pseudoobscura* show clearly that the introduced

TABLE 10.1

Dependence of the frequency of alleles on distance and time in D. pseudoobscura

	Total tests	Allelic combinations	Frequency of alleles
Within station	2068	44	0·0213
Between stations, within locality	2284	20	0·0088
Between localities	706	4	0·0057
Between regions	6294	26	0·0041
Within station (simultaneous)	594	15	0·0253
Within station (different times)	1474	29	0·0197
Between stations, within locality (simultaneous)	691	9	0·0130
Between stations, within locality (different times)	1593	11	0·0069

After Wright, Dobzhansky, and Hovanitz 1942

orange flies were rapidly eliminated. A number of experiments have been done in which mutant genes have been introduced into laboratory populations of *D. melanogaster*. Thus Carson (1961b) introduced females heterozygous for the recessive mutants sepia, rough, and spineless (together with the corresponding wild-type alleles) into populations that were otherwise homozygous wild-type or homozygous sepia, rough, and spineless. Rapid changes in gene frequencies occurred, but for each mutant the final frequencies in both types of introductions (single mutant to pure wild as compared with single wild into pure mutant) tended to be the same—that is, going towards an equilibrium point in each case. A sepia allele recovered from wild-type *D. melanogaster* captured in North Carolina was introduced into populations of various wild-type strains, and it was found that its frequency clearly depended on the source of the wild-type flies (Table 10.2)—that is, the result depended on the genetic background into which sepia was introduced (Wallace 1966c).

TABLE 10.2

Observed frequency of sepia homozygotes and the calculated frequency of the sepia alleles in four populations involving wild-type flies of different geographic origins

Source of wild-type flies	Number of individuals examined	Frequency of sepia homozygotes %	Gene frequency of sepia (%)
North Carolina	7731	0	0·4
Bogotá	8766	16·5	37·5
Barcelona	8925	2·8	20·2
California	8380	38·5	62·2

Each population consisted of 30 cultures (three sets of 10 each) maintained by the cross-transfer of adults each generation. The observations reported above were made one year after the cultures were started.   After Wallace 1966c

2. Dispersion and migration

The distribution of dispersal distances from a point is usually leptokurtic, with a maximum close to the point of origin, but with some individuals travelling great distances. This is the phenomenon referred to so far. However, on top of this are migrations from one area to another as in migrations to new continents, for example, that of the Negro from Africa to the United States. In *Drosophila*, the migration of flies to the Hawaiian Islands and the subsequent migrations between islands would be a good example (see Chapters 14–16). It is of interest, too, that in this latter case very few founder individuals are likely to be involved in many of the migrations (Carson 1970). To some workers, migration only includes this latter phenomenon and excludes the dispersion of genes as just discussed. However, if the definition of migration taken is *any movement of genes*, then it is clear that migration must include both phenomena, which from the geneticist's point of view is reasonable, as any movement of genes may have effects on the gene pool of the species (Parsons 1963b).

Migration, in the sense of large or small movements of immigrant groups in man, is therefore a phenomenon hardly studied in *Drosophila*, although it can, of course, be presumed to have occurred especially in the colonization of new habitats as in Hawaii. This would also be true of those species associated with man—e.g., *D. melanogaster*, *D. simulans*, *D. immigrans*, *D. bucksii*, and *D. hydei*—or of those specialized species spreading with the spread of a food host—for example, *D. buzzatii* which is associated with the cactus genus *Opuntia*, which originated in South America (Chapter 12) but is now widespread. In man, such large-scale movements can be documented with varying degrees of success. The genetic effects of these movements can be studied from the gene frequencies in the parent population, the original population in the new locality, and the hybrids, as has been done for the United States Negro, where it is estimated that the rate of migration from the white to the Negro population is from three to four per cent per generation (Glass and Li 1953). It is clear that *Drosophila* is a genus which so far is of rather less use than man in studying actual migration rates and distances, because in man these are frequently recorded by demographers. In man individuals can be followed, whereas in insects individuals must be marked and then captured. On the other hand, the fact that *Drosophila* is an insect does have advantages in experimentation.

The migrated genotype may interact with the new population. If differential migration occurs such that certain genotypes tend to migrate more than others, we have, following Fisher (1930), migrational selection, and this may occur as a result of environmental heterogeneity (see also Parsons 1963*b*). Fisher developed his arguments in terms of geographical distance. A likely conclusion, if there is a gradient of ecological conditions over a geographical transect, will be a gradient of fitnesses so leading to clines in gene frequencies, and the integrity of the cline may be maintained by migration. If active migration of genes continues between sections of a cline, there will be little tendency for parts of it to become isolated. *D. pseudoobscura* and other species show such geographical clines, and it must perhaps be concluded that this is aided by a degree of migration between adjacent populations.

Under natural conditions migration cannot therefore be overlooked; it is thus of interest to consider the performance of different genotypes in the laboratory where migration is possible, and this may be achieved using a migration-tube system such as is shown in Fig. 10.3. This allows individuals to move from one food chamber to another, and different strains show different powers of dispersion (Sakai, Narise, Hiraizumi, and Iyama 1958). Sakai *et al.* divided migration in *D. melanogaster* into two kinds: namely, random migration which occurs as a result of random movements of individual flies, and mass migration as a result of pressure of population density. Inter-strain differences for both activities were very pronounced, and it is assumed that the two types of activities are under genetic control. Furthermore, in a highly inbred and domesticated strain, the intensity of random migration was as low as 2·25 per cent of the total population per day, while in wild strains the figure was from 14 to 20 per cent. The low random-migration rate in the domesticated strain may well have been developed in response to a closed environment, where migratory ability would not be favoured. Narise (1968) found that while vestigial, *vg*, flies were eliminated in

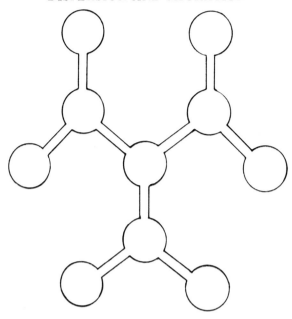

FIG. 10.3. Arrangement of ten migration-tubes. Flies were introduced initially into the central tube. (From Narise 1968.)

a population cage, an equilibrium value was obtained when migration was possible (Table 10.3), so indicating that migration can be an extremely powerful factor in the continuance of the species, or of different genotypes of the species. Furthermore, the actual migratory activity of vestigial flies is stimulated in the presence of wild-type flies, as assessed by estimates of migration six hours after mixtures of flies were introduced into the central tube of the migration-tube system. From later experiments, Narise (1969) concluded that the larger the difference between the genetic backgrounds of

TABLE 10.3

Percentage of vestigial flies in the population cage and the migration-tubes

	Generation						
	2	3	4	5	6	7	8
Cage	22·96	11·15	8·23	5·13	2·95	1·56	1·16
Tube	20·75	8·26	6·23	3·36	3·24	2·82	2·48
	Generation						
	9	10	12	14	16	18	20
Cage	1·29	0·87	0·67	0·37	0·40	0·29	0·25
Tube	1·80	1·29	1·41	1·02	0·87	1·06	1·05

After Narise 1968

the two strains in a mixture, the more intense the stimulatory effect on migratory activity appears. As different strains of the species *D. melanogaster* differ in rates of migration (vagility), it is not unreasonable that different species will differ in dispersal capabilities in the wild, as already discussed.

Population dispersion of *Drosophila* in the wild is therefore far from being understood, and considerable controversy exists regarding the data available. (The term dispersion is again used as this describes the work done so far in the wild in *Drosophila*.) Much work is based on the release and recapture of mutant eye varieties (Dobzhansky and Wright 1943). Traps are set in patterns allowing a mathematical analysis to be made of dispersal rates and distances from the number of flies caught. To what extent this can reflect what actually happens in wild *Drosophila* is debatable, since the individual genotype affects vagility (Narise 1968), and also the mutants are released from a crowded situation, which would tend to yield an overestimate of dispersion as population density is related to dispersal rate. Thus, the biological interpretation of dispersal rates—even if they can be described mathematically as has been done by Wallace (1966a)—is a matter of argument. Some of these problems can be dealt with by chemically labelling flies, so long as labelled individuals can be shown to be as equally vagile as unlabelled individuals. This is most effectively done using Richardson's (1969) technique of labelling flies in the wild by attracting them to a trap, so avoiding the release of laboratory-bred flies into the wild.

In conclusion, dispersion and migration are affected by factors such as population density, genotype, competition with other genotypes, the nature of the surrounding environment, and the living space of the species, but much more remains to be done. One factor which might increase the relative effectiveness of migration is the observation of Petit and Ehrman (1969), Ehrman (1969) and others in *D. pseudoobscura* and other species of *Drosophila*, that the mating success of certain genotypes of males increases as the males become rare relative to other males (see §7.2). If common, this density-dependent effect could substantially increase the effectiveness of migration, as migrant males would leave more than their proportionate share of offspring in localities where they were rare by virtue of being immigrant. On the other hand, some rare immigrants could well be eliminated in a foreign environment because of lowered general fitness in it.

11
POPULATION SIZE

1. Fitness and adaptedness

THE relation between population genetics and ecology culminates with considerations of population size. It involves a notion of population fitness, which has not been defined adequately in contrast with the relative fitnesses of individuals within populations. Since the ideal definition should include a prediction of future events, the idea of population fitness involves real problems in definition, more difficult than the concept of fitness of an individual. One definition of individual fitness is the capacity to leave adult fertile offspring, and so contribute to future generations. The relative Darwinian fitness of a genotype, or its adaptive value, is the average fitness of individuals of that genotype, or—if preferred—the average contribution which the carriers of a genotype make to the gene pool of the following generation relative to the contribution of other genotypes. Experimental difficulties arise because some components of fitness are difficult to measure. Thus, counting of survivors is frequently carried out, but fertility and fecundity, or the efficiency with which females obtain males, are less frequently assessed. The investigator thus often chooses those components of fitness that he can measure easily, and hopes that the unmeasured components will be correlated with what he has observed. Some complications in assessing fitness are obvious from Chapter 9, in which competition is considered. However, extending our concept of individual and genotypic fitness to population fitness involves us in a further order of complexity, as is apparent from the variable results in competition experiments according to the nature and proportions of competing genotypes. Even though defining population fitness seems difficult because of the uncertainty of future events, other things being equal, population size would seem to be related to population fitness. Thus, a genetic change that enables a population to cope with its surroundings more efficiently, and hence to increase its equilibrium size, could probably be regarded as a change improving the fitness of a population, or its adaptedness. Dobzhansky (1968) distinguishes between Darwinian or relative fitness, also called the adaptive value, and adaptedness. The adaptedness of an organism, a genotype, or a population to certain environments is a measure of the ability of the organism, or carriers of the genotype, or of the population to survive and reproduce in these environments.

Population geneticists often define the Darwinian fitness of a population as the weighted mean of the relative fitnesses of its genotypes. Frequently, the fitness of the genotype with the highest fitness is made equal to one, and the fitnesses of other genotypes are expressed as fractions of one. If the fitness, W_i, of the genotypes and their frequencies f_i are known, the fitness of the population is defined as

$$\bar{W} = \Sigma_i f_i W_i.$$

The fitness of the population so defined provides no information about the adaptedness of the population to its environment, since $\bar{W}$ is an average of relative numbers and does not involve over-all population sizes as above. This measure also does not provide for meaningful comparisons between populations. As Ayala (1969a) points out, Darwinian fitness is a relative, and adaptedness an absolute, measure.

The genetic load, further discussed in Chapter 13, refers to the fraction by which the population mean is changed as a consequence of differential fitnesses between genotypes. It can be defined (Crow 1958) as

$$\frac{W_{max} - \bar{W}}{W_{max}},$$

where W_{max} is the genotype with the highest fitness in the population. Clearly the genetic load provides no absolute information about the probability of survival or extinction in a population. If there is a balanced polymorphism due to heterozygote advantage, we can write the fitnesses of the three genotypes AA, Aa, and aa as $1-s$, 1, and $1-t$ respectively, where s and t lie between 0 and 1. If p and q are the frequencies of alleles A and a respectively, the fitness of the population is $\bar{W} = 1 - sp^2 - tq^2$ (see Appendix to Chapter 4). $\bar{W}$ will tend to increase under natural selection until an equilibrium is reached when $\frac{d\bar{W}}{dt} = 0$, from which $p_e = \frac{t}{s+t}$ and $q_e = \frac{s}{s+t}$, where p_e and q_e are the equilibrium gene frequencies. This result agrees with the proof given earlier (see Appendix to Chapter 4). At equilibrium, therefore:

$$\bar{W} = 1 - \frac{st}{s+t},$$

so that the genetic load of the population is $\frac{st}{s+t}$. A number of examples of polymorphisms for which the heterozygote is fitter than the two homozygotes have been discussed, and additional examples are presented in Chapter 12. Wright and Dobzhansky (1946), for example, found the relative fitnesses of the three karyotypes in *D. pseudoobscura*, ST/ST, ST/CH, and CH/CH to be about 0·7, 1, and 0·3 respectively at equilibrium at 25°C, giving $\bar{W} = 0·79$, so that the genetic load of the population is 0·21.

However, what happens if the population is monomorphic, and—say—is homozygous for the CH inversion? The fitness of this population is then

$\overline{W}=1$ and there is no genetic load. Introduction of the *ST* inversion would lead to $\overline{W}$ becoming <1, although the population contains two additional karyotypes *ST/CH* and *ST/ST*, both with *higher* fitnesses than *CH/CH*. This paradox arises from the calculation of population fitnesses as an average of relative numbers in the population, without any reference of the adaptedness of the population to the environment, and it provides no absolute information as to the fate of the population. The question therefore that arises is: how is the actual adaptedness of populations, as defined above, to be measured?

2. The innate capacity for increase

An operationally useful measure of the adaptedness of a population is the statistic r_m, or the innate capacity for increase in numbers, and is defined by Andrewartha and Birch (1954) as the maximum rate of increase attained by a population at a particular combination of quality of food, temperature, humidity, etc., when the quantity of food, space, and animals of the same species are kept at an optimum, and organisms of other species are excluded. Unfortunately, the estimation of r_m requires extremely laborious experiments, which have been performed with very few species, such as certain grain beetles and *Drosophila*. It is measured when the population has reached a stable age distribution—that is, a constant age schedule of births and deaths. The statistic r_m has been used successfully to compare the adaptedness of genetically different populations (Birch, Dobzhansky, Elliott, and Lewontin 1963, Dobzhansky, Lewontin, and Pavlovsky 1964). Thus they measured: the development time from egg to adult (F), egg, larval, and pupal survival (a), the proportion of adults still alive at age $x(l_x)$, and the number of eggs laid by a female on day x of her adult life (m_x).

From these estimates the expression

$$a \sum_0^t e^{-r_m(x+F)} l_x m_x = 1$$

is used to estimate r_m, where t=observed maximum life span.

Clearly, the experiments needed to estimate r_m are tedious in the extreme, which explains the few times such experiments have been carried out. An interesting derived value of r_m is the finite rate of increase

$$\lambda = \text{antilog}_e r_m,$$

which can be expressed as the rate of increase per day, per week, or per time interval. Small differences in r_m and λ can produce enormous differences in the number of animals within a fairly small number of generations.

In *D. birchii*, r_m tended to increase in experimental populations at both 20°C and 25°C for flies collected in the wild going from north to south from Rabaul in New Britain through Port Moresby (New Guinea) to Cairns (Australia), while for its sibling species *D. serrata*, r_m tended to decrease from north (Brisbane) to south (Sydney) (Birch et al. 1963). Not enough is known about the ecology of the two species to speculate on the possible adaptive significance of this pattern, but variations in r_m in different climates and laboratory environments must have general ecological significance, and are worth detailed study. In *D. pseudoobscura*, Dobzhansky, Lewontin, and Pavlovsky (1964) compared r_m in chromosomally monomorphic and polymorphic populations at two temperatures, 25°C and 16°C (Table 11.1). In

TABLE 11.1

Ordered comparisons of rates of increase r_m for different genotypes: the order is based on averages over the two years

At 25°C				At 16°C			
Genotype	1961	1962	Mean	Genotype	1961	1962	Mean
AR+CH	0·217	0·222	0·220	AR+PP	0·106	0·114	0·110
AR+PP	0·209	0·219	0·214	AR (Piñon)	0·108	0·107	0·108
AR (Mather)	0·198	0·215	0·207	AR+CH	0·104	0·096	0·100
AR (Piñon)	0·202	0·208	0·205	AR (Mather)	0·097	0·098	0·098
CH	0·200	0·185	0·192	CH	0·101	0·091	0·096
PP	0·154	0·187	0·170	PP	0·088	0·101	0·094

After Dobzhansky, Lewontin, and Pavlovsky 1964

general, r_m was superior in the polymorphic, as compared with the corresponding monomorphic populations, the superiority being more pronounced at 25°C as compared with 16°C. It was found that this superiority did not depend equally on all the components of r_m, but chiefly on a greater fecundity of females from the polymorphic populations. In most populations r_m increased in time over an 18-month period between 1961 and 1962, and this increase again depended chiefly on a higher fecundity, but also in some populations duration of life fell somewhat. Therefore, the adaptedness of the populations rose mainly through increasing fecundity. Unfortunately the r_m statistics, although providing measures of adaptedness, are limited in use, being sensitive to temperature, food (Ohba 1967), and other components of the physical environment, and in fact are only valid for the experimental environments and genetic types for which they are assessed. To make such estimates for natural populations would be a formidable task. Furthermore, r_m measures the innate capacity for increase of a population having no competitors of the same or of a different species, in spite of the fact that competition is one of the main forces in evolution. The problem of measuring adaptedness is, as Dobzhansky (1968) comments, evidently in need of further study.

3. Population sizes and productivity

Two related measures of adaptedness can be obtained, namely, 'productivity' and population size, which can both be assessed as population numbers or biomass. The two measures need not always be strictly correlated: an increase in the average longevity of the adults will lead to an increase in population size even if productivity remains constant; or an increase in productivity associated with a decrease in longevity may not lead to an increase of population size. These measures are only valid for comparisons between populations of the same species, or between populations of species which are closely related ecologically and phylogenetically. As such they are useful in *Drosophila*. Carson (1961*a*) detected differences in productivity and population size in experimental populations of *D. robusta* derived from different localities, such that a strain collected at the centre of the geographical distribution of the species performed better than a population collected at the margin of the geographical distribution (a discussion of central and marginal populations appears in §12.2). In *D. pseudoobscura*, population size and productivity were higher in populations polymorphic for two chromosomal rearrangements as compared with monomorphic populations (Beardmore, Dobzhansky, and Pavlovsky 1960, Dobzhansky and Pavlovsky 1961). A behaviourally polymorphic population of *D. willistoni* for pupation site showed a greater biomass than monomorphic ones (de Souza, da Cunha, and dos Santos 1970). The behavioural polymorphism involves the alternatives of larvae pupating on the food in a population cage, or on the bottom of the cage; the difference being controlled by a pair of major alleles at a locus. In *D. melanogaster*, the average population size of a strain carrying several mutant genes was about one-third that of a hybrid population of the mutant strain and an Oregon-R strain (Carson 1961*b*).

A method allowing the easy measurement of productivity and population size is the 'serial transfer' technique (Ayala 1965*a*). Adult flies are introduced into an experimental 'cage', usually a glass bottle, with a measured amount of food and allowed to lay eggs for a specified period of time, usually two or three days. At regular intervals the flies are transferred to new bottles with fresh food. When adult flies begin to emerge in the bottles where the eggs were laid, they are collected, counted, and weighed under anaesthesia, and then added to the bottle containing the adult population. The ovipositing adult flies are thus always in a single bottle with fresh food, while a number of bottles contain eggs, larvae, pupae, and newly hatched adults. The adult population is anaesthetized and counted at regular intervals. The technique also provides information about the number of flies emerging per unit food or per unit time. From the productivity or rate of emergence and the population size, estimates may be obtained of the average longevity of the adult flies.

Ayala (1965*b*, 1969*a*) has carried out a number of experiments using this technique. Thus, three strains of *D. serrata* were collected in Popondetta, New Guinea, Cooktown, Queensland, and Sydney, New South Wales, and two sets of populations were studied simultaneously for from fifteen to twenty generations at 25 °C and 19 °C (Table 11.2). The populations differed in their

TABLE 11.2

Mean population size and productivity with standard errors of six populations of D. serrata

Population	Temperature	Population size		Productivity per food unit	
		Number	Biomass (mg.)	Number	Biomass (mg.)
Sydney	25	1782±76	1274±53	550±17	329±10
Cooktown	25	2221±80	1486±49	568±20	334±13
Popondetta	25	1828±90	1185±62	477±13	251±8
Sydney	19	1803±87	1332±57	483±13	320±13
Cooktown	19	2017±84	1353±56	486±12	298±13
Popondetta	19	1580±52	1087±48	357±8	200±11

After Ayala 1969*a*

adaptedness to their experimental environments. The Cooktown strain had the largest population size at both temperatures; at 25 °C, however, the productivities of the Cooktown and Sydney strains did not differ significantly —that is, the Cooktown flies had a greater average longevity than the Sydney flies. Similarly, at 25 °C the productivity was significantly greater in the Sydney than in the Popondetta population, although these two populations did not differ significantly in population number. These considerations show, as mentioned already, that population size and productivity are not necessarily strictly correlated, and so the adaptedness of a population to a certain environment is better characterized by both parameters than by either alone. The adaptedness is also temperature-dependent, in that at 25 °C the Popondetta and Sydney population sizes were about equal, but at 19 °C the Sydney population size was larger; in other words, the Sydney flies are adapted to living in a climate considerably colder than Popondetta. *D. birchii*, a sibling species of *D. serrata*, was also studied, and it was found that there were striking differences between strains of the two species in adaptedness to experimental environments. Thus, the average population size of a Popondetta strain of *D. birchii* was 469±49 at 25 °C, and 428±33 flies at 19 °C, or about one-quarter of the Popondetta strain of *D. serrata*. In both species, hybrid populations between strains were generally larger and with higher productivity than the parents (Ayala 1966a). The data in Table 11.2 show that the central Cooktown *D. serrata* strains generally exceed the more northerly (Popondetta) and southerly (Sydney) populations for population sizes and productivity, which may be another central versus more marginal environment contrast, as for *D. robusta*.

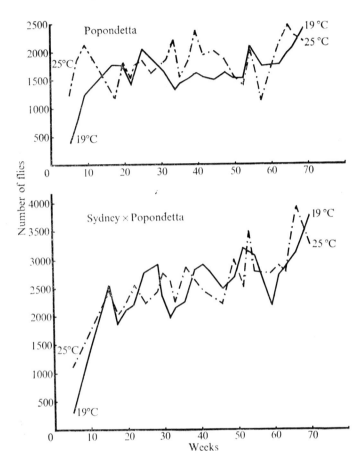

FIG. 11.1 Population size of two experimental populations, Popondetta, and Sydney × Popondetta, at 19°C and 25°C, in *D. serrata*. (From Ayala 1969a.)

Ayala (1965a, 1969a) studied two populations of the Popondetta strain of *D. serrata* at 25°C and 19°C for 70 weeks, or more than twenty-five generations. There is a trend towards increasing population size in both populations (Fig. 11.1), showing the evolution of adaptedness. Productivity also increased in both populations. However, in hybrid populations between the Sydney and Popondetta strains the evolution of adaptedness was more rapid, perhaps because of the greater variability of the hybrids (Fig. 11.1). Ayala (1966b, 1969b) has also shown that the new genetic variability induced by X-irradiation may increase the rate of adaptation to a new environment in both *D. serrata* and *D. birchii*. In these experiments, males were given 2000 r. of X-rays for each of three generations.

Ayala (1969a) considers that Fisher's (1930) fundamental theorem of natural selection may in theory be extended, at least in a qualitative form, to the evolution of the adaptedness of populations to certain environments. Fisher's fundamental theorem states that the rate of increase in fitness of a population at any time is equal to its genetic variance in fitness at that time. This theorem refers to the mean fitness of genotypes, and in the same way it could refer qualitatively to populations, perhaps in the form that the adaptedness of a population to the environments in which it lives increases at a rate proportional to the genetic variability present in the population. The results showing more rapid evolution of adaptedness when the genetic background is variable agree with this in a qualitative sense.

4. The physical environment

The preceding discussion shows the dependence of the size of a population on genetic factors and certain environmental factors; in this section environmental factors are considered more specifically. Temperature has been considered briefly already; for example, the innate capacity for increase of various geographic strains of *D. serrata* and *D. birchii* was found to be nearly double at 25°C as compared with 20°C (Birch et al. 1963), and *D. pseudoobscura* of various genotypic compositions had an average capacity for increase of more than double at 25°C as compared with 16°C (Dobzhansky, Lewontin, and Pavlovsky 1964).

There is considerable argument as to whether the amount of food is a check to numbers. For example, Andrewartha and Birch (1954) believe food limitation to be perhaps the least important among the possible ways by which population numbers can be limited, but Lack (1966) presents a contrary view based on studies almost exclusively on birds. Food quality is, of course, relevant, as is shown in §8.6. When a population is growing with a limited amount of food and space, the number of animals per unit of food and space gradually rises. Eventually an equilibrium must be reached when the number of births per unit time equals the number of deaths. The average population size need not, however, be proportional to the rate of births; for example, if in two populations there are the same number of births per unit time, one will be twice as large as the other if the average longevity of the animals in that population is also double that of the other. Using *D. serrata* in population cages of 400 c.c., Ayala (1968a) found that increasing the amount of food from two to three units of food per week led to the total population increasing by only 17 per cent, even though 50 per cent more flies were added per week. The increased amount of food caused an increase in the number of births which was associated with decreased longevity. It seems that population sizes under experimental conditions are controlled by food and space; and that as food is progressively increased, space becomes a limiting factor, the consequence of which is to reduce longevity. It is unlikely that living space is ever so limited for natural populations of *Drosophila*. The important point is that rates of births may not be strictly correlated with population size, and that factors other than food may exert a control over population numbers. Food is the major limiting factor, for an increase in food produces an increase in numbers, but since the increase in number is not proportional to the increase in food, space is postulated as the additional factor (see also Ayala 1966*a*).

In *D. serrata*, an experiment was carried out in which the amount of space as well as the amount of food was varied (Ayala 1968*a, b*). An increase or decrease in the amount of food of 33 per cent produced an increase or decrease of 26 per cent in population size. A change of 33 per cent in the amount of space resulted in a 10 per cent change in population size. Again both food and space are limiting factors, although within the range studied food played a greater role than space. Comparable results have been obtained in *D. birchii, D. melanogaster*, and *D. pseudoobscura*.

Therefore population sizes are affected by the genotype and environmental factors such as temperature, food, and space. Of major importance is the genotype, and furthermore the rate of evolution of adaptedness to a new environment is positively correlated with the initial amount of genetic variability in the population. The effect of temperature is clear, indicating how climatic variables are likely to affect population numbers, and the same presumably applies for humidity. Finally, food and a place to live may jointly limit the maximum size that a population may reach. The real problem that emerges is the extreme difficulty of extrapolating these experiments to the wild, since a trait like population size, while having a meaning in the laboratory, may have little meaning in the wild. This is because population size and the more general concept of the adaptedness of the population, must depend on all the variables discussed in this book, and many others as well, being one manifestation of their interaction. We have not yet considered the role of the biotic components of the environment, such as the effects of closely related species exploiting a limited food resource, although the types of problems involved can be seen from competition experiments between different genotypes within species. These issues are considered further in §14.3.

12
DISTRIBUTION OF GENOTYPES WITHIN SPECIES

1. Inversion polymorphisms and natural selection

MANY species of the Diptera have inversion polymorphisms. It seems likely that inversion polymorphisms can become more easily established in the Diptera than in many other groups because of the absence of crossing-over in the males, and the selective elimination of dicentric and acentric fragments during meiosis in females, such that the balanced non-cross-over gametes pass selectively to the egg nucleus (Sturtevant and Beadle 1936). Furthermore, crossing-over may be suppressed within the inversion. Inversion polymorphisms frequently occur in other insects, such as grasshoppers (White 1956, 1962), but have been very thoroughly studied in *Drosophila* (Dobzhansky 1951, 1957*a*, *b*, da Cunha 1955). In *D. willistoni* as many as fifty different inversions have been found in nature, such that in central Brazil an individual is heterozygous for an average of nine inversions (see Dobzhansky 1957*b*, 1959). Presumably the inversions are maintained in the population by natural selection.

Convincing evidence for the selective control of inversion frequencies comes from observations on the change in frequency of some inversions after their introduction into population cages. In *D. pseudoobscura*, two sequences, Standard (*ST*) and Chiricahua (*CH*), gave stable equilibria of about 70 per cent *ST* and 30 per cent *CH*, irrespective of the initial inversion frequencies at 25°C (Wright and Dobzhansky 1946). Other pairs of inversions also give characteristic equilibria. Stable equilibria under the relatively constant environment of the population cage can be most simply explained by an advantage of the inversion heterozygotes over the homozygotes (see Appendix to Chapter 4). In population cages it is possible to estimate the fitnesses of the three genotypes by using the changes in gene frequencies in relation to to the number of elapsed generations (Wright and Dobzhansky 1946). Some examples are given in Table 12.1(a), and all show heterokaryotype advantage for inversions from the Californian localities Piñon Flats and Mather, which are in the San Jacinto mountains of the Sierra Nevada ranges. At 16·5°C, however, little change occurred in frequencies in the population cages, showing the dependence of the equilibrium on the environment, in this case temperature. At the intermediate temperature of 22°C, it is of interest that in some cases polymorphism occurs, and in some cases monomorphism for

TABLE 12.1
Relative fitnesses of individuals homozygous and heterozygous for various gene arrangements

(a) Within localities (from gene frequency changes related to number of generations)

	Arrangement		Fitnesses		
	1	2	1/1	1/2	2/2
Piñon Flats	ST	CH	0·85	1·00	0·58
Piñon Flats	ST	AR	0·81	1·00	0·50
Piñon Flats	AR	CH	0·86	1·00	0·48
Mather	ST	CH	0·78	1·00	0·28
Mather	ST	AR	0·64	1·00	0·58
Mather	AR	CH	0·81	1·00	0·60

(b) Between localities (from the F_2 of inter-population hybrids between Mather (M) and Piñon Flats (P))

Population number	Arrangement		Fitnesses		
	1	2	1/1	1/2	2/2
45	AR^M	CH^P	1·28	1·00	0·47
46	ST^P	AR^M	0·63	1·00	1·51
47	ST^M	AR^P	1·31	1·00	0·57
48	ST^M	CH^P	1·18	1·00	0·48
49	ST^P	CH^M	1·38	1·00	0·43

From data of Dobzhansky and colleagues summarized by Wallace 1968

population cages containing Arrowhead (*AR*) and *CH* (Van Valen, Levine, and Beardmore 1962).

Considering the 25°C cages, the F_2 between *ST* and *CH* and some other pairs of polymorphic inversions show an excess of heterozygotes (Dobzhansky 1947*a*). Samples of eggs were deposited on food in the population cages and then the larvae were grown under optimal conditions. Agreement with Hardy-Weinberg proportions was good, although there was a very slight excess of heterozygotes. However, in the adult flies from the cages there was a very much greater excess of heterozygotes. Thus selective elimination of the homozygotes must occur during the larval stages under the high level of competition in the cages (Dobzhansky 1947*a*). The same general conclusion was reached in the South American species *D. pavani*, based on competition experiments in vials (Budnik, Brncic, and Koref-Santibañez 1971).

At 25°C, Beardmore et al. (1960) compared population cages polymorphic for *AR* and *CH* chromosomes with cages monomorphic for these chromosomes. They found that (1) the polymorphic populations produced more individuals than the monomorphic ones, (2) the number of individuals hatching per egg-laying cup was less variable in the polymorphic populations, (3) the mean weights of individuals were similar, but the variances of weights of individuals were lower from the polymorphic populations, (4) there was a greater biomass from the polymorphic populations, and (5) wing lengths were similar, but wing-length variances and asymmetry were lower from the polymorphic populations. Thus, the polymorphic populations were found to be more productive. The likely interpretation of the lower variability and wing asymmetry of the polymorphic populations is that they have superior developmental homeostasis—that is, they are better buffered against environmental variations and so vary less. In these experiments, larval crowding was high and the adults were relatively uncrowded. Further experiments (Dobzhansky and Pavlovsky 1961) have shown that the polymorphic populations were superior to the monomorphic ones when the adults were crowded and the larvae relatively uncrowded.

Beardmore et al. (1960) carried out their experiments at 25°C, but they have been repeated at 16°C (Battaglia and Smith 1961). At 16°C the heterokaryotypes *AR/CH* were not heterotic as they were at 25°C, as would be expected from the population cage experiments (Wright and Dobzhansky 1946, Dobzhansky 1947b, 1948a) where fitnesses of karyotypes were much more similar at 16·5°C as compared with 25°C. There was some slight evidence that the polymorphic population was superior in fitness to the monomorphic ones, based mainly on the facts that more individuals were produced from the polymorphic population, and that asymmetry of wings was greater in the monomorphic populations. It is to be expected that different environments will give different results, since the fitnesses of genotypes must be a function of their precise environment. (See also §11.2, where the innate capacity for increase was shown to be superior in polymorphic as compared with monomorphic populations, especially at 25°C.)

The greater fitness of polymorphic *Drosophila* populations as compared with monomorphic is considered by Beardmore (1963) to be evidence for mutual facilitation between genotypes, and that the polymorphic population allows of more ready adaptation to a variety of ecological niches than do the monomorphic populations. For example, populations of D. *pseudoobscura* polymorphic for *AR* and *CH* were superior to monomorphic populations *AR* or *CH* when they competed for the available food resources with another species, D. *serrata* (Ayala 1969b). Mutual facilitation has been shown experimentally between strains of D. *melanogaster* for larval viability, perhaps due to diffusible metabolic products produced by the larvae in the food medium from one genotype affecting another (Dawood and Strickberger

1969). Modifications of fitness due to interactions between genotypes, the exact result not necessarily being predictable based on the fitnesses of the genotypes tested in isolation, have been shown by others, for example, Lewontin (1955) in *D. melanogaster* and Lewontin and Matsuo (1963) in *D. bucksii* (see Chapter 9). These results have analogies with the polymorphic situation in that the different genotypes may have slightly different ecological requirements, so that mutual facilitation (or interference) would result. Beardmore (1961) showed that *D. melanogaster* populations of common ancestry had a higher additive genetic variance in a diurnally varying temperature condition than in a constant temperature. He interpreted this as being due to a greater variety of niches in the fluctuating environment. Long (1970) extended this type of result by beginning with twelve populations of *D. melanogaster* derived from a wild population, maintained for two years in laboratory culture. The populations were then divided into four groups, one of which was kept at a constant temperature of 25°C, and the others under temperature cycles fluctuating between 20°C and 30°C with periods of one day, one month, and three months respectively. Productivity, competitive ability, and egg-to-adult survival under NaCl- and $CaCl_2$-stress, were measured. In an index combining the results of all tests, over-all population fitness was found to be greater under more frequent environmental variation. The results therefore support the idea that natural selection under a variable environment tends to result in greater population fitness than does natural selection in a more uniform environment. Additional evidence with a general discussion is given by Beardmore (1970). The recent experiment in *D. willistoni* (de Souza *et al.* 1970) has lent support to the idea that polymorphic populations are fitter than monomorphic by exploiting more than one ecological niche for a polymorphism involving a behavioural trait—namely, whether larvae prefer to pupate on the food in a population cage, or on the bottom of the cage—the difference being controlled by a pair of major genes at a locus. Polymorphic populations with both types of flies reach larger population sizes and have a greater biomass than monomorphic populations. That the polymorphism depends on the multi-niche environment is shown by the rapid elimination of genes of preference for pupation outside the food in populations kept in 250-cm.^{3} bottles.

It is not therefore surprising that studies on natural populations of *D. pseudoobscura* (Dobzhansky 1947*b, c*) have shown annual cyclical changes in inversion frequencies. Under natural conditions at Piñon Flats, California, the *ST* arrangement is about twice as frequent as *CH* during the winter when little reproduction occurs. During the spring *ST* declines and *CH* increases in frequency, so that by the beginning of June *CH* is more frequent than *ST* (Fig. 12.1). With the onset of warm summer weather these trends are reversed, and by October *ST* again becomes twice as common as *CH*. Laboratory experiments have provided some possible explanations. In population cages with *ST* and *CH* kept at a relatively warm temperature (25°C), in which reproduction is permitted in an uncontrolled fashion, so that the cages become crowded with flies, an initially small proportion of *ST* flies increased in frequency until the *ST* chromosome frequency reached 70 per cent and then remained constant. For cages kept at 16·5°C no change in chromosome frequency occurred, as expected from results already given. When bottles at 25°C were inoculated with a small sample of flies and, by repeated transfers, the number of larvae and adults was kept down to fifty per bottle, *CH* persisted at a slightly higher frequency than *ST*. This may well duplicate the likely low level of competition present at Piñon Flats in the

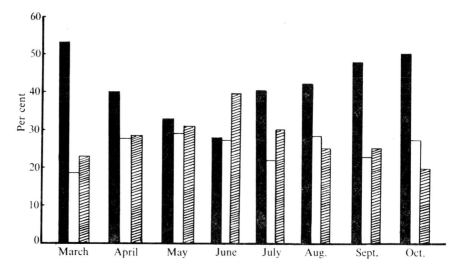

FIG. 12.1. Relative frequencies of chromosomes with different gene arrangements in different months in the population of Piñon Flats. Black columns—*ST* chromosomes; white columns—*AR* chromosomes; hatched columns—*CH* chromosomes. (From Dobzhansky 1947*c*.)

spring, when there would be a period of rapid population expansion. In other words, the explanation of the observed seasonal changes probably lies in the differential adaptation of the inversion sequences to regular changes in temperature and competition each year (Birch 1955). The observed regular altitudinal changes in inversion frequencies, as described by Dobzhansky (1948*b*) in *D. pseudoobscura* (Table 12.2), are then not unexpected. Thus the polymorphism is an efficient way of accommodating regular environmental variations and other types of environmental heterogeneity, by giving the population greater powers to adapt to a multiplicity of environments. Similarly, geographical clines have been often reported in North America (see, for example, Dobzhansky 1950, Mayhew, Kato, Ball, and Epling 1966).

In other species such as the South American *D. flavopilosa* (Brncic 1962), altitudinal clines have been reported which are associated with fitness differences between karyotypes for two temperatures at which the larvae were permitted to develop in the laboratory (Brncic 1968). On the other hand, in *D. pavani* Brncic (1969) did not find seasonal or geographical variations. Thus, in some species the chromosomal polymorphism is flexible and leads to varying frequencies of inversions in different environments due to variations in selective values of homo- and heterokaryotypes, whereas in other species the chromosomal polymorphism is more rigid, presumably because of less variation of selective values. Examples of rigid species, apart from those already mentioned, include *D. subobscura* and *D. willistoni*; examples of

TABLE 12.2

The incidence of third chromosomes with different inversions, ST, AR, and CH in populations of D. pseudoobscura, *which live at different elevations in the Sierra Nevada Mountains of California*

Elevation (feet)	Inversion frequency (%)			
	ST	AR	CH	Others
850	46	25	16	13
3 000	41	35	14	10
4 600	32	37	19	12
6 200	26	44	16	14
8 000	14	45	27	14
8 600	11	55	22	12
10 000	10	50	20	20

After Dobzhansky 1950

flexible species include *D. robusta* (see Krimbas 1967, Brncic 1969). Another factor which may be relevant in explaining the distribution of the karyotypes may be the presence or absence of different bacteria and yeasts found in the wild in the intestinal canal. In *D. pseudoobscura* for example, da Cunha (1951) found that the relative fitnesses of ST/ST, ST/CH, and CH/CH varied according to the micro-organism supplied to population cages.

In *D. funebris*, a further flexible species, Dubinin and Tiniakov (1945, 1946a–c, 1947) found that inversion heterozygotes were more frequent in the populations of cities than in country populations. The city of Moscow showed a high level of polymorphism, and furthermore larger cities and towns showed more polymorphism than smaller ones. The incidence of the inversions underwent cyclic seasonal changes, in Moscow the frequency being lowest in winter and early spring (March) and apparently trebling by late summer and autumn (August–September). By exposing flies to low temperatures, $+3°C$ and $-2°C$, Dubinin and Tiniakov hypothesized that during the warm period certain inversions are favoured, and in the cold season this trend is reversed; also the cold-resistant genotypes are more fecund than the cold-susceptible ones. There is clearly a situation here of considerable interest from the ecological genetic point of view.

In *D. pseudoobscura*, of interest so far as natural environments are concerned are the observations of Dobzhansky, Anderson, Pavlovsky, Spassky, and Wills (1964) who reported changes in the relative frequencies of the third chromosomes with different gene arrangements over the period between 1940 and 1963 in California. In most localities the changes occurred in the same direction—that is, *ST* increased and *CH* decreased—although in some localities they took place earlier or more rapidly than in others (see also Dobzhansky 1947c). Similar changes seem to have occurred throughout the Pacific States of the U.S.A. (Dobzhansky, Anderson, and Pavlovsky 1966). The causes are obscure, but Dobzhansky *et al*. (1964) considered the possibility that the massive use of pesticides in Californian agriculture was a contributing factor, or perhaps the rapid spread of new gene complexes may be a cause. Oshima and Watanabe (1966) showed that *ST* seems fitter than *AR* when exposed to traces of the insecticides DDT and dieldrin, and this in fact parallels the changes in the American south-west. On the other hand, Anderson, Oshima, Watanabe, Dobzhansky, and Pavlovsky (1968) found no evidence that the genetic changes observed in natural populations may have been caused by the insecticides DDT and dieldrin.

While these studies have been primarily based on laboratory experimentation, attempts have been made to correlate variations in chromosomal arrangement frequency with environmental factors faced by the natural populations in the field. Work on laboratory populations alone, with subsequent extrapolation to the field, is not possible without the introduction of possible sources of error, as no laboratory 'environment' as yet constructed can simulate macro- and perhaps more importantly micro-fluctuations encountered by a natural population. Furthermore, changes in the arrangement frequency can be an extremely rapid process, whether seasonal (Strickberger and Wills 1966, Dobzhansky 1947b), altitudinal (Stalker and Carson 1948), or geographical variation (Carson 1958a) is considered, which would suggest that a given arrangement is best adapted to a given environmental situation. Laboratory experimentation does, however, permit the discovery of factors likely to be of importance in the wild.

2. Coadaptation for inversion polymorphisms

Dobzhansky (1950) compared pairs of arrangements from different ecogeographic populations under experimental conditions in *D. pseudoobscura*. On the whole, heterozygote superiority broke down completely. Table 12.1(b) (see page 137) shows relative fitnesses for a number of geographically mixed populations between Mather and Piñon Flats. The fitnesses are based on deviations from the 1:2:1 ratio expected in the F_2 offspring of inter-population hybrids. Fitnesses were measured in this way because changes in inversion frequencies in mixed populations are erratic, so that estimating fitnesses from observed frequency changes in population cages was not possible. Such one-generation fitness tests measure the differential survival of larvae (which can be termed more strictly viability), which is one component of fitness. In every experimental population without exception one homozygote exceeded the heterozygote in fitness. This indicates that the gene arrangements within the inversions are unique to specific geographic populations and are mutually adjusted. Dobzhansky (1950) called this mutual adjustment of gene complexes within the same population to give high heterozygote fitness 'coadaptation'. Clearly the gene complexes from different populations will not have had the same opportunity for mutual adjustment of their genic contents—that is, they would not have been selected for coadaptation, so that heterozygosity for inversion sequences from different localities will not be expected to lead to increased fitness.

The mutual adjustment of gene complexes occurring within populations must have evolved by natural selection in the environments of the populations in question. Dobzhansky and Levene (1951) have demonstrated the development of heterozygote superiority in population cages for pairs of inversion sequences derived from different localities which did not at first show any superiority. They took ST from Piñon Flats, California, and CH from Mexico. In the F_2 of the inter-racial cross, fitnesses such that $ST/ST > ST/CH > CH/CH$ were indicated. In a population cage it would be expected that ST would be fixed. However, after some 14 months an equilibrium of 69 per cent ST and 31 per cent CH was obtained with fitnesses becoming $ST/CH > ST/ST > CH/CH$, showing that the hybrid California/Mexico population had developed overdominance during the course of the experiment. Thus, heterozygote fitness was selectively modified by natural selection operating on the increased genetic variance generated by segregation and recombination in the hybrid population. In general, when the gene complexes are disrupted by recombination, overdominance may disappear, as observed also by Dobzhansky and Pavlovsky (1953, 1958) who found frequent evidence for a lowering of heterozygote fitness in population cages begun with flies heterozygous for inversions from different localities in *D. pseudoobscura*, *D. paulistorum*, and *D. willistoni*. The cages with flies from different localities soon showed a preponderance of homozygotes, while flies from the same locality maintained a high frequency of heterozygotes as coadapted complexes. Even so, as already pointed out, overdominance occasionally develops during the course of the experiment for flies heterozygous for inversions from different localities, and it is difficult to predict when this may occur (Dobzhansky and Pavlovsky 1953).

Levene, Pavlovsky, and Dobzhansky (1954) studied experimental populations of *D. pseudoobscura* containing all three of the arrangements *ST*, *AR*, and *CH*, and showed that their results could not be explained by a model assuming constant fitnesses. It was expected from the known fitnesses that *CH* would eventually be eliminated and that a balanced polymorphism of *ST* and *AR* would remain (see Bodmer and Parsons 1960). However, *CH* was not eliminated, and a reasonable fit to the observed equilibrium frequencies showed an increase of *ST/CH* fitness, with a minor change in *AR/AR* and *CH/CH* fitnesses. The increase in the fitness of the heterozygote *ST/CH* may have occurred through the selection of modifiers which interact with the genic contents of the inversions, or by selection of inversions carrying altered genic contents which may have arisen by recombination. Thus, we have another case of the possible modification of fitnesses by natural selection under experimental conditions, and analogous phenomena may be expected in the wild.

In a theoretical study, Haldane (1957) examined the necessary conditions for the establishment of new inversion sequences: he concluded that there must be a general trend towards overdominance. Similar conclusions have been reached by Bodmer and Parsons (1960, 1962). A trend towards overdominance as found in the experimental situations discussed could occur by the incorporation of modifiers which must interact with the existing gene contents of the inversion sequences. In this way the overdominance of each particular inversion heterozygote will depend on the population in which it is, since in different environments different modifiers will be incorporated. Therefore, since different modifiers are likely to be incorporated in different populations it is reasonable that in between population crosses, overdominance will frequently be broken down as observed.

3. Marginal and central populations

In certain species of *Drosophila*, notably *D. willistoni* and *D. robusta*, inversion heterozygosity decreases rapidly as the margin of the species range is approached; in other words, central populations may be better adapted to their diverse environments by means of a high level of inversion heterozygosity. It has been suggested that in marginal regions there is a greater need for more flexibility to adapt to new environmental conditions. The low level of inversion heterozygosity permits a higher level of recombination, thus providing novel gene combinations more rapidly than in populations containing high frequencies of inversions (see, for example, Dobzhansky, Burla, and da Cunha 1950, da Cunha and Dobzhansky 1954, da Cunha, Dobzhansky, Pavlovsky, and Spassky 1959). Marginal populations may thus be better able to respond to evolutionary opportunities, and can probably form new species more readily than populations from the centre of the distribution of the species. In *D. robusta*, central populations have from seven to nine inversions, while peripheral populations have from one to six inversions (Carson 1958*a*, *b*). Low levels of polymorphism also occur in island populations of *D. willistoni*, and populations on small islands tend to be less polymorphic than on large islands (Dobzhansky 1957*b*). The amount of polymorphism in a population therefore tends to be related to ecological diversity and the environmental opportunities which the population exploits. Prevosti (1966) has given similar evidence for *D. subobscura* in Western Mediterranean populations. It should, however, be pointed out that diversity is not necessarily a *direct* function of the size of the environment (MacArthur 1965), since if the environment is patchy, with a large number of microhabitats, more genetic diversity might be expected as compared with a uniform environment occupying a similar area. In §15.2, this point is further discussed when species distributions are considered. Thus, with this proviso, the above comments on marginal and central populations may be valid.

Cosmopolitan species such as *D. melanogaster* are relatively free from natural chromosomal polymorphisms (*D. robusta* and *D. willistoni* are restricted to parts of North and tropical South America respectively). The genetic system of *D. melanogaster* thus resembles the marginal populations of *D. robusta* and *D. willistoni*, and so should be able to respond rapidly to new conditions. *D. pseudoobscura* seems to be intermediate in this respect, as inversion heterozygosity is a feature of the species, but it is restricted to the third chromosome. Restriction of inversions to one chromosome seems to occur more frequently than would be expected by chance in species with small numbers of inversions. Perhaps this is because the organism compensates for the low rate of recombination in the chromosome with inversions by a high rate in other chromosomes (Schultz and Redfield 1951). Free recombination in chromosomes without inversions permits flexibility for adaptation to new environments, whereas the inversion sequences are probably adapted to a given series of predictable, but often widely varying, environments. Inversion heterozygosity for one chromosome is therefore an example of the way in which natural selection may accommodate the opposing evolutionary requirements of fitness in one environment and flexibility to adapt to others. This argument may be an oversimplification, because other methods of varying over-all recombination exist, such as chiasma localization and chromosome number variations (see Darlington 1937, White 1958, 1962), although within a genus such as *Drosophila*, where inversions assume so much importance, the argument presented is probably relevant.

In examples cited in this section we are discussing not what environmental variables are involved, but the consequences of the whole environment on a species and the genotype so developed. Appropriate laboratory experiments may allow a clearer definition of marginal versus central environments but, as already pointed out, such experiments are liable to error because of the impossibility of simulating natural environments, which have a level of unpredictability difficult to create under experimental conditions. A start has, however, been made, since Band (1963) compared the frequency of drastic variants (lethals and semi-lethals having a fitness up to 10 and 50 per cent respectively of normal as homozygotes) in *D. melanogaster* populations collected from South Amherst, Massachusetts in three environments—a constant temperature 25°C, a fluctuating environment with a wide range 14°C between the daily maximum and minimum which approximates to the wide-range temperatures observed in the natural environment during the pre-collection periods, and a fluctuating environment with a narrower range 8°C between daily maximum and minimum which approximates to the narrow-range temperatures observed in the natural environment. In all cases the average over-all temperature was 25°C. It was found that the frequency of lethals and semi-lethals was higher for the narrow-range temperature than for the wide-range temperature or for the constant environment. Thus, extreme environments lead to a reduction of genetic diversity, as does a constant environment. The latter observation accords with evidence of Beardmore (1961) already cited, who found a higher additive genetic variance in *D. melanogaster* in a narrow-range diurnally-varying temperature condition $25°\pm5°C$ than in a constant environment. Similarly, Oshima (1969) found for $25°\pm5°C$ with a mean of 25°C, compared with a constant 25°C, that fitnesses of heterozygotes for lethal, semi-lethal, and quasi-normal genes (those which are not lethal or semi-lethal as homozygotes) were greater in the fluctuating environment. Thus, greater genetic diversity may be associated with favourable environments which may comprise fluctuations without being too extreme. More work needs to be done in this area, especially as the interpretation of the data is not easy, since the amplitude of the fluctuations seems to be of importance and could vary in its effect both within and between species (see §8.1).

Hoenigsberg, Castro, Granobles, and Idrobo (1969) compared *D. melanogaster* populations in European (Hungarian) and neo-tropical (Colombian) environments. They found that the frequencies of drastics (lethals plus semi-lethals) were far higher in Colombia (about 35 per cent) than in Hungary (about 10 per cent). The Hungarian ecosystem is such that the population size of *D. melanogaster* undergoes serious periodic shrinkage, whereas the neotropical areas can support very large populations all the year round. Other published data cited by Hoenigsberg *et al.* indicate that in general the more northern populations have lower frequencies of drastics. Such variations, if associated with the ecosystem, lead us also to ask about the possibility of seasonal shifts, and although some work has been reported it is a little early to discuss it in detail (see Hoenigsberg, Granobles, and Castro 1969). Ives

(1970) has made a start for populations from South Amherst, Massachusetts, where winter temperatures are severe, and over-wintering probably occurs as larvae in a continuously available rotten-apple pile he studied (see §8.1). On emergence in June and July, many of the adults contained chromosomes with lethals allelic to other lethal chromosomes, presumably because they were derived from a few founder individuals. However, later in the season the proportion of individuals carrying lethal chromosomes allelic to others fell, and the proportion of non-allelic lethal chromosomes increased, presumably by the mixing of populations, and from mutations giving new lethals.

Similarly, marginal populations of *D. willistoni* tend to contain fewer drastic variants than do populations at the centre part of the species distribution (Cordeiro, Townsend, Petersen, and Jaeger 1958; Townsend 1952). In *D. robusta* the genetic load, estimated by comparing the egg-to-adult lethality in inbred and out-bred crosses of F_1 lines derived from wild flies collected during consecutive years, was found to be more or less uniform in the central populations from year to year, so perhaps reflecting the homeostatic properties characteristic of a large population. In marginal populations there was a considerable difference in the loads between years (Nair 1969).

Dobzhansky, Hunter, Pavlovsky, Spassky, and Wallace (1963) have made similar observations in *D. pseudoobscura*, where a marginal population in the Andean highlands around Bogotá, separated from the main body of the species by 1500 miles, was found to have the lowest proportion of lethals and semi-lethals of any population of *D. pseudoobscura* studied. Conversely the heaviest loads were found in the most prosperous (and central) populations. United States populations which are comparable with the Colombian populations come from the Death Valley and Mount San Jacinto regions of California; here we are probably dealing with ecological rather than geographical marginality. The Death Valley populations have to cope with extreme summer aridity and, at high elevations in the mountains, with long and severe winters. These populations are presumably reduced to very few survivors in unfavourable seasons so allowing periodic inbreeding, drift, and hence a general reduction in the genetic load. To a lesser extent this is true of Mount San Jacinto and the Colombian highlands. Using twenty-four assayable and (it was hoped) randomly chosen loci for enzymes and larval proteins, Prakash, Lewontin, and Hubby (1969) have shown that the isolated Bogotá population has only about one-third of the variability shown by the main body of the species, and shows the greatest amount of genic divergence of all populations.

In *D. melanogaster*, Band and Ives (1968) found that lethal and semi-lethal frequencies were significantly positively correlated with total rainfall in summer. In one period, 1947–64, the annual rainfall per summer was almost two inches less than in the 1938–46 period, and the frequencies of lethals and semi-lethals were lower in the later than in the earlier period. Thus, the drier environment, which is probably but cannot be proven to be more marginal, is associated with lower frequencies of lethals and semi-lethals, although, as shown in §8.1, desiccation is a stress to which *Drosophila* is susceptible, which agrees with this hypothesis.

In conclusion, greater genetic diversity is associated with central environments, implying diversity of ecological niches, favourable temperatures, or other climatic conditions; while in marginal environments, implying limited ecological niches with unfavourable temperatures or other climatic conditions leading periodically to very small population sizes, there is less genetic diversity. Put succinctly, the amount of genetic variability in a population is a function of the diversity of the environment in which the population lives. Beardmore (1970) has also argued from his experimental work in *D. melanogaster* that the amount of heterogeneity of the habitat is generally correlated with the amount of variability in gene pools. However, genetic variability as a function of the environment does *not* always imply a causal relationship, since reduced variability can be a function of reduced population size, as has already been discussed. It is therefore clear that the interpretation of much of the experimental data on marginal and central populations poses problems of peculiar difficulty, and the story is still far from complete.

4. The integrated genotype within and between species

The frequencies of lethal and semi-lethal chromosomes are probably higher in the more widespread and versatile species, such as *D. pseudoobscura* and *D. willistoni* (see, for example, Krimbas 1959) than in others. Another manifestation of the ecological versatility of these two species, as compared with *D. persimilis* and *D. prosaltans* for example, is that the former two have been shown to release more variability by recombination than the latter two. The differences in the amount of variability released by recombination presumably reflect a difference in the amount of concealed genetic variability in these species; the higher genetic variability occurs in the ecologically most versatile species (Spassky, Spassky, Levene, and Dobzhansky 1958; Spiess 1958, 1959; Dobzhansky, Levene, Spassky, and Spassky 1959; Levene 1959; Krimbas 1961). This result stresses the integrated nature of the genotype, its dependence on the species, and its ecology. Such studies are, however, difficult to interpret at the between-species level, since within species genetic architectures can vary between central and marginal populations, degrees of inbreeding, population-size fluctuations, etc., as discussed in §12.3.

Within species, studies have made the concept of a single wild-type genotype untenable. Instead, the details of the integrated genotype must be regarded as a characteristic of local populations rather than of the whole species, so that there is a multiplicity of wild-type genotypes each adapted to local conditions. Considerable evidence for this has already been cited, such as co-adaptation for inversion heterokaryotypes within populations (§12.2) but some other experiments are discussed here. By crossing strains of flies obtained from the same and different geographic localities, Vetukhiv (1953, 1954, 1956, 1957) studied the survival of larvae of *D. pseudoobscura* and *D. willistoni* under crowded conditions. Two other components of fitness, egg production, and longevity were also studied. Table 12.3 gives the results of the F_1 and F_2

TABLE 12.3

Viability, fecundity, and longevity of intra-population and inter-population hybrids of D. pseudoobscura *and* D. willistoni

	F_1	F_2
Survival		
D. pseudoobscura		
Intra-population	1·00	1·00
Inter-population	1·18	0·83
D. willistoni		
Intra-population	1·00	1·00
Inter-population	1·14	0·90
Eggs/female/day		
D. pseudoobscura		
Intra-population	1·00	1·00
Inter-population	1·27	0·94
Longevity		
D. pseudoobscura		
Intra-population	1·00	1·00
Inter-population (16°C)	1·25	0·94
Inter-population (25°C)	1·13	0·78–0·95

After Vetukhiv 1953, 1954, 1956, 1957, and see Wallace 1968

generations, as summarized in Wallace (1968). In each case, the observation for the intra-population F_1 and F_2 has been set to 1·00 to provide a standard for comparison. The F_1 inter-locality hybrids in all cases consistently have higher larval survival, egg production, and greater longevity than the intra-locality hybrids. At first sight it is a little difficult to explain the high fitness of the inter-locality F_1s, but it seems that the greater hybridity might account for part of the improvement, as there would be greater average heterozygosity in the inter-locality F_1s, so that they would reveal the effects of fewer deleterious genes. This increased inter-locality fitness of the F_1s is lost in the F_2 generation, in other words, following recombination, the F_2 inter-population hybrids lose all the advantages of the F_1 hybrids. Therefore, the highly integrated genotypes evolved within populations are broken down in between population crosses. This argues for coadaptation of gene complexes as a general characteristic of local populations. Similar observations have been made in *D. melanogaster* (Wallace and Vetukhiv 1955), and for body size in natural populations of *D. pseudoobscura* from Canada to Mexico (Anderson 1968). These and other results (Brncic 1954, 1961; Wigan, 1944; and see Bodmer and Parsons 1962) make the concept of a single wild-type genotype of a species untenable.

A recent study of Anderson (1969) confirmed this for the selection coefficients for lethal heterozygotes in data of Dobzhansky and Spassky (1968). His analysis showed that in a genetic background from their own populations, five out of forty-five lethal chromosomes were statistically significantly heterotic and none were significantly deleterious, whereas in foreign genetic backgrounds about ten of forty-five lethal chromosomes were significantly deleterious and none were significantly heterotic. In other words, it seems that heterosis evolved for lethals in their own populations is lost when the lethals are put into a foreign genetic environment. Crow (1968) discussed the possibility of interaction between viability effects of pairs of chromosomes as homozygotes, and gave methods of estimating it. Kosuda (1971) used his method to compare interactions between natural second- and third-chromosome homozygotes of *D. melanogaster*, and artificial combinations of second and third chromosomes isolated from different wild flies and combined together with appropriate matings. A coefficient of interaction $I < 1$ means that positive synergistic interaction occurs, or that the viability of the double homozygote is smaller than the product of the two single homozygote viabilities. For the natural pairs, $I = 0.925 \pm 0.025$, which exceeds that of the artificial pairs, for which $I = 0.898 \pm 0.042$. Although these means are not significantly different, the variances are ($P < 0.01$), showing more variability in artificial combinations. Furthermore, only the artificial pairs give any inter-chromosomal synthetic lethals. The evidence suggests the possibility of natural selection in natural pairs not leading to such extremes as in artificial combinations of pairs, and thus argues for interactions between, as well as within, chromosomes being under the control of natural selection. More evidence is clearly necessary.

The results just described show that it may be of importance in the future to carry out detailed studies of genes (and polygenes) differentiating local populations. On the other hand, in *D. subobscura*, McFarquhar and Robertson (1963) did not find any evidence for differentiation into local races as assessed by F_2 data. While this may be so, and may be a feature of the genetic system of *D. subobscura*, the evidence advanced above for other species does seem convincing. Thus it seems that the gene pools of some species are more prone to evolving coadapted gene complexes, such that the gene pool of *D. pseudoobscura* can change, readily evolving geographic races each characterized by its own coadapted gene complexes, whereas the gene pool of *D. subobscura* is probably more rigid.

In conclusion, the studies referred to in this chapter generally show the genotype to be integrated within species at one level, and within populations at another, evolving specific genetic architectures in response to the environments prevailing upon species and populations. Although the environment has hardly been discussed specifically, if we are to assume that it is a relevant factor in the evolution of the integrated genotype, studies of environment in relation to genotype, or more generally local populations, are of crucial importance.

13
ENVIRONMENTS, ENZYME VARIANTS, AND GENETIC ARCHITECTURES

1. Extreme environment heterosis

CERTAIN environmental stresses have been discussed in some detail. From the population point of view, the population cage experiments of Wright and Dobzhansky (1946) and Dobzhansky (1947a, 1948b) in *D. pseudoobscura* show large differences in fitness between heterokaryotypes and homokaryotypes leading to heterokaryotype advantage and a stable polymorphism at 25°C and not 16·5°C, with an intermediate situation at 22°C (Van Valen, Levine, and Beardmore 1962 and see §12.1). There is also a tendency for there to be greater differences between polymorphic and monomorphic populations for r_m, the innate capacity for increase in *D. pseudoobscura* at 25°C as compared with 16°C (see Table 11.1, page 130). For *D. pseudoobscura*, 25°C represents a more extreme environment than 16·5°C. Other examples of heterokaryotype advantage in extreme environments already cited for *D. pseudoobscura* include desiccation (Thomson 1971), cold-tolerance (Heuts 1948), and mating speed and duration of copulation at high temperatures (Parsons and Kaul 1966). Thus it could be argued that in more optimal environments heterokaryotype advantage is minimal, and is only expressed in extreme environments. If this is a general phenomenon, it would provide a further explanation, in addition to those already advanced in the literature, for the association between a high level of genetic polymorphism in natural populations and a not excessive genetic load.

A definition of the genetic load appeared in §11.1, with particular reference to that associated with a polymorphism maintained by heterozygote advantage (referred to as the segregation or balanced load). Any factor that influences gene frequencies can lead to a change in the average fitness, and hence produce a genetic load. Additional factors producing a genetic load include mutation, recombination, heterogeneous environment, meiotic drive or gamete selection, maternal-foetal incompatibility, finite population size, and migration (see Crow and Kimura 1970). The genetic load concept came from Haldane (1937), who discussed the mutational and segregation loads, but it first came into real notice as a result of a paper by Muller (1950) in relation to the load of mutations in man. Wallace's (1968) view is that in this sense the term is useful, but does not warrant the rigorous mathematical treatment it has received. Work in the past few years has shown that the average out-breeding species is highly polymorphic; for example, Lewontin and Hubby (1966) concluded that the average individual in *D. pseudoobscura* is heterozygous for at least 12 per cent of genes in the entire genome based on studies of enzyme and protein variants. Furthermore, the proportion of polymorphic loci to total loci has been conservatively estimated as from 30 to 70 per cent. The possibility that this is due to heterozygote advantage was examined by Lewontin and Hubby, with the conclusion that the consequent genetic load would be enormous. Since then, various explanations of this dilemma have appeared, such as density-dependent selection (§7.2 and §9.3). Sved, Reed, and Bodmer (1967) and others have considered the problem and shown that the assumption of heterozygote advantage for many loci leading to an enormous genetic load is not necessarily true. This is because, although individuals with optimal genotypes are of primary importance in determining the genetic load as usually defined (§11.1), they are sufficiently rare in the population to play little or no part in determining the average selective advantage at individual loci. In other words, these optimal individuals with fitness W_{max} would in theory be complete multiple heterozygotes, which in a highly polymorphic species would not exist.

To return to the situation under discussion, in more optimal environments the species might have a high level of polymorphism without a concomitant genetic load, but in times of environmental stress heterozygote advantage would occur for some polymorphisms according to the stress. Essentially a situation can be imagined where a population is subjected to a number of environmental stresses during its existence, throughout which heterozygote advantage for certain loci or inversions might be favoured according to the stress, so facilitating more ready adaptation to the stress. Since a high level of heterozygote advantage would be restricted more or less to stress environments, which would probably affect a minor part of the population, the over-all genetic load would not be high, given the assumption of a multiplicity of extreme ecological niches in which different heterozygotes would be advantageous.

Analogous to this are the several observations of heterosis for hybrids between inbred strains under extreme environments in this case high temperatures: for example, larval survival (Parsons 1959b, and see Fig. 13.1) and longevity (Parsons 1966) in *D. melanogaster*, and growth rates in *Arabidopsis thaliana* (Langridge 1962) and maize (McWilliam and Griffing 1965); and low temperatures: for example, viability in *D. melanogaster* (Fontdevila 1970) and growth rates in maize (McWilliam and Griffing 1965). Furthermore, a similar phenomenon has been found for the effects of homozygotes of chromosome II compared with the corresponding heterozygotes of *D. pseudoobscura* with respect to viability at 25°C (Dobzhansky, Pavlovsky, Spassky, and Spassky 1955) and for cold-temperature resistance (Marinkovic, Crumpacker, and Salceda 1969); also for pre-adult survival for chromosome

II in *D. melanogaster* at 29°C (Tobari 1966). These are not natural population situations, but are analogous in that for a normally out-bred species, homozygotes and heterozygotes are being compared and show that the genetic load, in this case the inbreeding load, is higher in adverse environments than in favourable ones. Therefore, the more severe the physical environment, the greater the level of heterosis (and genetic load) developed (see Parsons 1971 for examples in other species).

So far as temperature-dependent heterosis is concerned, Langridge (1968) has proposed an explanation at the molecular level—namely, that heat-sensitive enzymes are the most common consequences of mutations that do

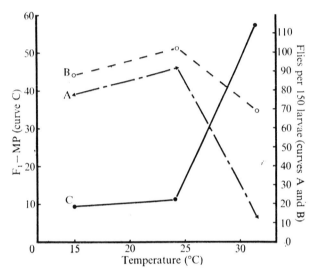

FIG. 13.1. Heterosis at three temperatures for larval survival in *D. melanogaster* (plotted by Langridge 1962 from data of Parsons 1959*b*). A = means of the mid-parent (MP); B = means of all F_1s; C = B–A.

not inactivate the enzyme, that some of these mutations are expressed in the organism only at high temperatures, and that complete dominance of the normal phenotype is expected in a heterozygote. Langridge has found some support for this hypothesis for mutations affecting the structural gene for β-galactosidase in the bacterium *Escherichia coli* K_{12}. A recent report of Gibson and Miklovich (1971) on alcohol dehydrogenase in *D. melanogaster* shows in some cases that enzyme activity is significantly heterotic after incubation of homogenates for 10 minutes at 40°C, as compared with prior incubation at 25°C. This mechanism could account for desiccation-dependent heterosis, if a reasonably high correlation can be assumed between the two stresses as found in *D. melanogaster* (Parsons 1970*a*). So far as cold temperatures are concerned, low-temperature-sensitive enzymes seem relatively rare (Langridge 1968); thus, perhaps heterosis might be expected to be less extreme at low temperatures, but it does exist as discussed above.

An alternative hypothesis to explain heterosis in extreme environments comes from evidence for the coadaptation of gene complexes between chromosomes in heterozygotes, which is broken down in homozygotes. Because homozygotes are less commonly subjected to natural selection than heterozygotes in out-breeding species, their fitnesses will normally be lower and probably this can be expected to be particularly marked in extreme environments. Because heterozygotes are the genotypes most commonly exposed to natural selection in out-bred species, the general observation that they are less variable in a multiplicity of environments is then not unexpected—and as a concomitant of this is the increased level of heterosis found in stress environments. Needless to say, neither of these hypotheses is exclusive of the other since the latter one could be put into molecular terms. Even if the molecular hypothesis is too restrictive, it is likely that the major environmental stresses to which a wild population of *Drosophila* is likely to be subject will be associated with variations in temperature and humidity. The term extreme-environment heterosis is adopted as it encompasses all stresses yet still admits the importance of temperature (and humidity).

2. Enzyme variants and genetic loads

In recent years there has been a spectacular increase in analysing individuals of many species (see Stone, Kojima, and Johnson 1969) for enzyme and protein variants by gel electrophoresis, which allows a quick survey to be made of size and electric charge differences in the protein molecules in different individuals—the differences reflecting altered molecular structures. As already discussed, such studies have revealed that a large fraction of loci controlling enzyme- and protein-type specificity are polymorphic, and this has led to considerable discussion as to how the population maintains this level of polymorphism. There is, in fact, little experimental evidence for overdominance for enzyme polymorphisms except for a brief report by Richmond and Powell (1970) for a sex-linked enzyme locus, tetrazolium oxidase in *D. paulistorum*. However, assuming the existence of environment-dependent heterosis, overdominance may not be expected for most experiments which would normally have been carried out in non-extreme environments. On the other hand, frequency-dependent selection has been shown to be responsible for the maintenance of biochemical polymorphism in at least one system, *Est-6* in *D. melanogaster* (Kojima and Yarbrough 1967), and this could well turn out to be increasingly important in the maintenance of polymorphisms (see Wallace 1968 for a further discussion), especially as at equilibrium the associated genetic load would be minimal (see §7.2 and §9.3).

It may also be that linkage disequilibria are important in the maintenance of gene-enzyme polymorphisms (Sved 1968). It is desirable briefly to explain the term 'linkage disequilibrium'. In the case of two linked loci, each with two alleles A, a, and B, b, it is expected that if there are no selective differences between genotypes, the frequencies of the gametes AB, Ab, aB, and ab will gradually approach the equilibrium values given by the product of the frequencies of the two alleles contained in a gamete. Thus, if p_1, q_1, and p_2, q_2, represent the gene frequencies of genes A, a, and B, b, respectively, the equilibrium frequencies of the gametes AB, Ab, aB, and ab are expected to be $p_1 p_2$, $p_1 q_2$, $q_1 p_2$, and $q_1 q_2$ respectively, a situation which may be called a 'gene frequency equilibrium' (Bodmer and Parsons 1962). Fisher (1930) described a model of interacting fitnesses leading to an excess of either coupling (AB and ab) or repulsion (Ab and aB) gametes over that expected from gene frequency equilibria. A quantity, D, referred to as the *linkage disequilibrium*, has been used to quantify this discrepancy, and can be written as $g_{AB} g_{ab} - g_{Ab} g_{aB}$, where g_{AB}, g_{ab}, g_{Ab}, and g_{aB} are the frequencies of gametes AB, ab, Ab,

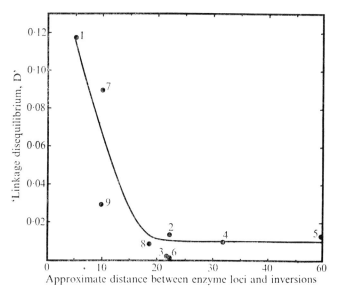

FIG. 13.2. The values of linkage disequilibria D between a small chromosomal segment marked by a locus producing enzymes and an inversion on the same chromosome in *D. melanogaster*. The data are for flies collected in the wild from Katsunuma, Japan. The points on the graph represent the following enzyme-inversion combinations:

(1) *ADH–In 2LB*, (2) *ADH–In 2RC*,
(3) *Est-6–In 3RB*, (4) *Est-6–In 3RH*,
(5) α *GPD–In 2RC*, (6) *Est-C–In 3RI*,
(7) *ODH–In 3RG*, (8) *ODH–In 3RH*,
(9) *Est-C–In 3RG*.

(From Kojima, Gillespie, and Tobari 1970.)

and aB respectively. The magnitude of D has been shown to depend on the fitnesses of the genotypes involved, and on the magnitude of the recombination fraction between them. A substantial literature has been built up examining such relationships, but a good recent review is given by Kojima and Lewontin (1970), where linkage disequilibria are discussed fairly extensively in relation to the evolutionary role of linkage and epistasis.

A linkage disequilibrium was reported in a population cage between alleles at the leucine amino-peptidase D (*Lap-D*), and acid phosphatase-1 (*Acph*-1) loci in *D. melanogaster* (O'Brien, MacIntyre, and Fine 1968) which are 3 map-units apart. Kojima, Gillespie, and Tobari (1970) found substantial linkage disequilibria in *D. melanogaster* for enzyme loci and inversions, which are 10 map-units or less apart (Fig 13.2). There was a dependence of D on the recombination fraction such that $D \to 0$ as the recombination fraction increases.

Franklin and Lewontin (1970) have shown the theoretical possibility that almost neutral loci may show strong linkage disequilibria over considerable map distances, and that higher-order interactions among loci in multi-locus systems increase in importance as the number of loci increases. They make the general point that the theory of population genetics should not deal with gene frequencies at individual loci, but with whole chromosomes, their recombination properties, and the effect of homozygosity of segments of various length. According to them, such considerations when fully developed may lead to a complete rethinking of population genetics including the abandoning of the concept of the genetic load, which is mainly developed in terms of individual loci.

3. The balanced genotype

It should, however, be pointed out that Mather's (1943) concept of the balanced genotype also refers more to the effects of individual chromosomes than of individual loci. Mather (1943) discussed how genes affecting a metrical trait might be arranged on the chromosome, assuming stabilizing selection. Representing a gene increasing a metrical trait by $+$, and by $-$ if it decreases it, and considering two loci, gametes $+-$ and $-+$ will be less extreme than $++$ and $--$. The gametes $+-$ and $-+$ are then *well balanced* with respect to high and low genes and will be favoured over $++$ and $--$, which give extreme values and will tend to be eliminated by stabilizing selection. At this level, therefore, we are talking about arrangements of genes along a chromosome, and it can be argued that under certain assumptions for many loci, arrangements $+-+-+-\ldots$ or $-+-+-+\ldots$ would be favoured by natural selection (see Lee and Parsons 1968 for further details). Now if there are no interactions between the genes at the two loci, the coupling double heterozygotes $++/--$ and the repulsion double heterozygotes $+-/-+$

will give intermediate phenotypes and hence optimum fitness. These are the only two genotypes affected by linkage between the two loci, as all other genotypes for the two loci will be homozygous at one locus at least. However, the repulsion gametes +− and −+ are well-balanced compared with the coupling gametes ++ and −−, so that natural selection will tend to preserve the repulsion gametes rather than the coupling gametes. If the loci are linked, a repulsion double heterozygote will produce balanced gametes +− and −+ in excess of unbalanced gametes ++ and −−, and the reverse will occur from a coupling double heterozygote. Therefore, for linked loci with a recombination fraction less than 50 per cent the average fitness of offspring from the repulsion double heterozygotes will exceed that from coupling double heterozygotes, and any mechanism preserving repulsion gametes at the expense of coupling gametes will be favoured by natural selection. Selection for tighter linkage to preserve repulsion gametes is the obvious mechanism. In other words, intermediate genotypes with adjacent genes in repulsion arrangements will be preserved by natural selection's adjusting linkage relations between the loci to accommodate the opposing evolutionary requirements of fitness in a given environment and flexibility to adapt to a new environment. The existence of variation of the genotype controlled by these polygenes implies that some individuals will depart from the optimum and show reduced fitness. Variation to this extent is disadvantageous, but it is the type of variation likely to be exploited during environmental changes. Thus Mather's concept involves the recombination properties of whole chromosomes, as is also put forward by Franklin and Lewontin (1970).

Out-breeding promotes heterozygosity; consequently a heterozygous species will evolve a balanced polygenic system adapted to heterozygosity. Therefore, there is expected to be a balance between chromosomes as well as within chromosomes. This means that many out-bred species will be expected to be polymorphic for many genes. If an out-breeding species is forced to inbreed, the original polygenic balance will be disrupted, exposing genotypes not previously subject to natural selection. A great deal of evidence in out-breeding species for the poorer fitness of homozygotes compared with heterozygotes has been found in various *Drosophila* species (see Lee and Parsons 1968). The poor viability of homozygotes in a normally out-bred species shows that the species have balance between chromosomes, which is designated *relational balance*, a term similar in meaning to coadaptation, which was initially developed for gene complexes within inversions (§12.2). Conversely, those species which inbreed as a rule and are more homozygous must evolve balance within chromosomes, which is termed *internal balance*, although many inbreeding species have evolved mechanisms for preserving heterozygosity. In this way the theory of the balanced genotype leads us to consider the consequences of homozygosity of whole or parts of chromosomes rather than specific loci, as also discussed by Franklin and Lewontin (1970).

Therefore, in conclusion, Mather's (1943) concept of the balanced genotype poses problems similar in type to those put forward by Franklin and Lewontin (1970), although in a far less rigorous way mathematically. Furthermore, just as Franklin and Lewontin's (1970) models show large linkage disequilibria, the same applies to fitnesses corresponding to Mather's balanced genotype, where in fact under certain circumstances small disequilibria are obtained for recombination fractions of 50 per cent, but needless to say maximal disequilibria are obtained for small recombination fractions (Parsons 1963c). Another point is that, given fitnesses corresponding to Mather's concept of balance, the linkage disequilibria are probably generally larger when the repulsion double heterozygote viability exceeds the coupling double heterozygote viability. This provides a further way whereby, in theory, balanced repulsion gametes would be favoured over coupling gametes (Parsons 1963d). In fact, it might be predicted that if repulsion double heterozygotes are in general favoured, then the repulsion double heterozygote viability would be higher than the coupling double heterozygote viability, since the repulsion double heterozygote would be exposed more frequently to natural selection. Mather's ideas were put forward before realization of the extremely high level of polymorphism in out-breeding species, and the consequent discussion of genetic loads. To reiterate, these complexities make it difficult to measure or to assess the meaning of the term genetic load in natural populations, since they imply that fitnesses as measured in experimental situations involve blocks of genes, usually whole chromosomes, and hence the complexities implied by this.

4. Enzyme variants and the environment

It is likely that the next few years will provide a great deal of information for selective differentials in different environments for enzyme and protein variants. It has been argued that in *D. ananassae* there is a cline for the major allele in the *Est-C* system in the Pacific, such that its frequency increases from north to south towards the equator, and then south of the equator decreases again (Stone *et al.* 1969). However, this interpretation has been obscured by later data (Johnson 1971). Data were from Samoa, Fiji, Ponape, Guam, Yap, Palau, Philippines, Majuro and Hawaii. The conclusion was that allelic frequencies for four loci with a moderately high level of polymorphism tended to differ more between geographically remote localities than between localities situated close to one another. This does not preclude a dependence on environment, even though the islands which are mostly very isolated from each other do not generally vary in any systematic way with regard to environmental factors, but rather show irregular diversity in geography, climate, and biota. The abrupt allelic frequency changes are argued to be consistent with adaptation in view of the type of environmental heterogeneity present. In

D. melanogaster, Berger (1971) found the allelic frequencies of five dimorphic loci to show a general similarity from different collecting sites in the United States. No marked seasonal differences were found. However, in the same species, Beardmore (1970) has presented experimental data in the laboratory showing that the frequency of alleles at the esterase-6 locus depends on temperature, such that the esterase-6^F allele tends to be at a higher frequency in experimental populations at high temperatures. An example of a possible association with environment comes from the harvester ant, *Pogonomyrmex barbatus*, from Texas (Johnson, Schaffer, Gillaspy, and Rockwood 1969) for two esterases and a malic dehydrogenase. Principal component analysis was used to compare the patterns of variability of the allelic frequencies with environmental factors (plant life areas, elevation, rainfall, January and July temperatures, and growing season). Significant correlation was particularly evident with respect to weather and the pattern of variability in both esterases, from which it may be inferred that natural selection is important in determining the allele frequency patterns. Overdominance was not found, but it was concluded that all the enzyme systems studied were affected by the environmental factors. The species *Pogonomyrmex barbatus* was chosen because of its abundance over much of Texas, and because sampling could be done so as to avoid most migration difficulties. Furthermore, compared with *Drosophila* in Texas, populations are stable and are maintained at relatively high densities throughout the year, and individuals live considerably longer than most *Drosophila* species. An example of a dependence on temperature comes from the freshwater fish *Catostomus clarkii* in the United States where the frequency of alleles for polymorphic serum esterases varies with latitude (Koehn and Rasmussen 1967, Koehn 1969). The activity of the allele more frequent in southern populations increases as temperatures increase from 0° to 37°C, whereas the activity of the allele more frequent in northern populations increases as temperature decreases. In *D. melanogaster*, Gibson (1970) has shown that ethanol supplementation in the diet affects gene frequencies at a locus for alcohol dehydrogenase. This may be of some ecological importance because there is probably considerable variation in the amounts of ethanol produced by yeast fermentation in *Drosophila* habitats. Finally, Beardmore (1970) has shown that the frequency of the esterase-6^F allele is correlated with population sizes in the laboratory, but it is difficult to relate this to natural populations at this stage.

In the butterfly, *Hemiargus isola*, a dimeric esterase shows rich electrophoretically-detectable variation in natural populations in Texas (Burns and Johnson 1971). This enzyme variation is controlled by multiple alleles at an autosomal locus (*Es-d*), and in each population sampled there are from nine to fourteen alleles. Two of these, *Es-d*^{100} and *Es-d*^{80}, predominate, forming about 80 per cent of the gene pool. The frequencies of these two alleles are notably stable, being about 65 and 19 per cent respectively, in ecological and variable space and time. The rare alleles collectively account for the remainder, and individually occur at frequencies varying from 0·2 to 5 per cent, the exception being $Es\text{-}d^{115}$, which is scarce in all localities except one where it has a frequency of 10 per cent based on a small sample size. Burns and Johnson argue that environmental heterogeneity is a prime factor maintaining the polymorphism, and that the near uniformity of the frequencies of the two common alleles suggests in this case that the selective forces do not relate to obvious environmental factors.

O'Brien and MacIntyre (1969) found an average level of heterozygosity of 22·7 per cent in eight populations of *D. melanogaster* for ten gene-enzyme systems analysed by starch-gel electrophoresis, and it is of interest that an experimental population maintained at an average breeding size of 5000 individuals for twenty years showed levels of polymorphism equivalent to those of natural populations. Two populations of *D. simulans* had much less variability. A similar conclusion was reached by Berger (1970) for six enzyme loci in five populations containing the two species sympatrically. It is, of course, difficult to say whether the greater range of environments (temperatures) tolerated by *D. melanogaster*, as compared with *D. simulans*, is associated with these variations in levels of polymorphism (§8.1), but it is worth considering. Studies of central and marginal populations within species would also be worth while. As reported in §12.3, Prakash *et al.* (1969), using enzyme and larval protein variants, found that an isolated population of *D. pseudoobscura* had only about one-third of the variability shown by the main body of the species.

Finally, as we have seen, much evidence shows that various inversions in some *Drosophila* species are overdominant within populations. Prakash and Lewontin (1968) have shown considerable genetic divergence between inversions of *D. pseudoobscura* for enzyme and protein variants, and the available evidence suggests that such genic differences date from the origin of the inversions, predating the present distribution of the species. Even though no direct studies of environmental variables were carried out, there is the possibility that loci controlling enzyme and protein variants associated with different inversions could be of importance in explaining the associations of inversions with season, altitude, and environmental factors. The likelihood of this is enhanced if a molecular basis, either partly or wholly, is accepted for environment-dependent heterosis. Information on this is likely in the near future, and some of the definitive work is very likely to involve comparisons of enzyme and protein variants under extreme and more optimal environments. Therefore, it is critical to define clearly the environments used, especially in laboratory experiments, since results under different environments could well lead to different interpretations.

PART III

14

DISTRIBUTION DATA BETWEEN SPECIES

1. Species distributions

THERE are a number of publications showing variations in population numbers and relative proportions of different species of *Drosophila* collected in the wild. In some cases the variations have only been marginally correlated with environmental changes. Dobzhansky and Pavan (1950) described local and seasonal variations in various *Drosophila* species in thirty-five localities in seventeen bioclimatic regions of Brazil, and found some species to form nuclei with high and low population densities, and some to be distributed relatively more uniformly. Food preference differences and seasonal differences were also found. Their work suggests many problems of an ecological-genetic nature, most of which are only soluble by detailed work.

Cyclical large population expansions in a uniformly tropical wet climate (Moen Island, Truk, eastern Caroline Islands) in *D. ananassae, D. hypocausta*, and *D. anuda*, are believed to be explainable by a dependence upon the fruiting of certain types of fruit-trees, such as breadfruit. *D. melanogaster*, never an abundant species on Moen Island, on the other hand did not undergo any expansions during the period of collection, and so presumably was independent of the fruiting of any major plant species in this environment (Pipkin 1953). In a further survey (Pipkin 1965), the influence of feeding and breeding habits upon population size was investigated in seventy-three species of neotropical forest-dwelling ground-feeding Drosophilidae in central Panama. Some species preferred small drier fruits and blossoms for both adult and larval feeding, and entered traps to a limited extent. Because of their feeding and breeding habits, they underwent population expansions where such occur in wet seasons, when fruits and blossoms would suffer less from desiccation. Those species using fleshy fruits entered traps more readily, and some tended to undergo population expansions in the dry seasons when they could use fleshy fruits. Those species using both feeding sites also tended to undergo population expansions in the dry season. In other words, population size variations are dependent on the ecology of the species in question.

In the same way, Cooper and Dobzhansky's (1956) data on the natural species of *Drosophila* inhabiting the Sierra Nevada Mountains in the Yosemite region of California showed differentiation of a number of species with respect to environmental variations in space and time. Some of the species occurred at all elevations at which collections were made (850–11 000 feet), but were relatively more abundant in some altitudinal zones. Other species were more narrowly specialized, not occurring at all altitudes. Within localities, temporal variations were found on both a seasonal, and a year-to-year basis. Among other observations, it was found that *D. persimilis* was commoner at high elevations than its sibling species *D. pseudoobscura*. Phaff, Miller, Recca, Shifrine, and Mrak (1956) found variations in the yeast flora in the alimentary canal of the species of adult *Drosophila* in the Yosemite region, but several species of *Drosophila* contained the same common species of yeast. The micro-organisms found in the alimentary canal of adult *Drosophila* differed from those found in places where the larvae are known to feed, indicating that the larval and adult yeast foods of these species differ, and that adult flies do not regularly feed from their breeding sites (Carson, Knapp, and Phaff 1956). It was also shown that there are behavioural mechanisms causing adults of different species to be attracted preferentially to different species of yeast isolated in the Yosemite region, although all the yeasts tested attracted all *Drosophila* species tested to some extent (Dobzhansky, Cooper, Phaff, Knapp, and Carson 1956). da Cunha, Dobzhansky, and Sokoloff (1951) also found that yeasts isolated from the crops of *D. persimilis* and *D. azteca* near Mather, California showed differential attractiveness to the two species. *Drosophila*-yeast relationships could be of considerable importance in explaining distributions in different communities. Dobzhansky *et al.* (1956) considered fourteen 'natural' species in the Yosemite region, although there are another four species found there—namely, *D. melanogaster. D. immigrans, D. bucksii,* and *D. hydei*—which are associated with man, and so do not belong to the 'natural' community. Greater numbers of species of *Drosophila* are found in other regions, especially the tropics, where the variety of micro-organisms and substrates is likely to be higher. As the variety of foods and of consumers increases, so the specialization of food preferences may be expected to increase. This is likely in the *Drosophila*-yeast relationships in parts of Brazil (Dobzhansky and da Cunha 1955), and in the Hawaiian species discussed below.

The accumulated data on the Hawaiian species of the Drosophilidae show that thirty-two endemic and six introduced genera of plants belonging to thirty families serve as ovipositional sites, and of these the genera *Cheirodendron* (Araliaceae) and *Clermontia* (Lobeliaceae) serve as ovipositional and larval food sites for more than half the species that have been reared from all sources (see Heed 1968 and Carson *et al.* 1970). It is considered by Heed (1968) that these two plant genera account for much of the adaptive radiation of the Drosophilidae in the Hawaiian Islands. The natural food substrates of the adults are less well known than are those of the larval stages, but adults of various species have been observed engaging in feeding behaviour in the field, for example, on the fermenting leaves and bark of *Cheirodendron*, the leaves, flowers, fruit, and bark of Lobelicids, the leaves of *Pisonia*, various fungi, and the exuding sap from the recently cut stumps of large fern-trees (*Cibotium*). Adults do not randomly distribute themselves throughout the forest but tend to accumulate in relatively small habitats which suit their ecological requirements, the main controlling factors being probably wind intensity, humidity, temperature, light intensity, food sources, and acceptable ovipositional sites. Thus, moderate wind currents and light intensities, humidities below 90 per cent, and temperatures above 21·1 °C seem generally to be avoided. During periods of heavy overcast weather when relative humidities approach 100 per cent, and especially if misty rain is falling, the flies tend to move upwards into the vegetation and can be found on the under-surfaces of leaves and limbs up to 6 to 10 feet above the ground. However, on cloudless sunny days when humidity falls the flies rapidly disappear, presumably seeking out small poorly-lighted areas where the humidity is high and the light intensity is low. Temperature is, however, probably the most important factor, since most of the species, both of drosophiloids and scaptomyzoids, are found at about 1000 feet of elevation, reaching a maximum at 3000–4000 feet. Significantly, species of *Clermontia* and *Cheirodendron* that serve as larval substrates and adult food sources for many species are rarely found below 1000 feet and are most abundant in the 3000- to 4000-foot range.

Carson *et al.* (1970) consider that the major factors responsible for the evolution of the extraordinary numbers of species on the Hawaiian Islands reside in the spartan nature of the food supply, which resulted in the evolution of a low reproductive rate and consequently small population sizes. The native forests, in fact, have an extreme paucity of fleshy fruits that are so important for many species of *Drosophila* in other parts of the world. Added to this are the probably infrequent migrations between adjacent islands which act as effective isolating barriers, the isolated habitats on specific islands due to volcanic and meteorological activity, the evolution of specialized behaviour patterns (Chapter 15), and the invasion of specialized food sources such as the leaves of a number of plants and the eggs of spiders.

Detailed work on habitat has been done on the South American species *D. flavopilosa*. Females lay their eggs only in the flowers of *Cestrum parqui* (Solanaceae), and their developmental cycle takes place within these flowers (Brncic 1966). By counting the number of flowers of *C. parqui* containing pre-adult forms of the fly in different geographic regions, it is possible to estimate the abundance of the species in relation to climate, available food, and other insects which inhabit the same flowers as competitors, predators, or parasites. The main factors regulating population density seem to be the climate, a species of Thysanoptera, *Thrips tabaci* (as a competitor), and two species of parasitic Hymenoptera. Brncic (1970) has reported two other species closely associated with single plant species—namely, *D. allei* associated with *Datura arbustiva* (Solanaceae), and *D. appendiculata* associated with *Chusquea* (Bambuseae). It is not possible to breed these three species in the laboratory with the usual methods employed for cosmopolitan species, and this feature seems to be a product of their ecological restriction. Three or four other species of *Drosophila* have been reported which feed on a single species of flower (Pipkin, Rodríguez, and Léon 1966) based on collections in Panama and Colombia in neotropical forests. In general, monophagous flower-feeding *Drosophila* occupy plants with long blossoming periods of 4–9 months. Other species studied in these regions were found to be polyphagous using a variety of host plants which generally have a short blossoming period of 1–3 months, the most versatile having a wide range of distribution and include the cosmopolitan *D. bucksii*. A number of polyphagous flower-feeding *Drosophila* species can be cultivated on laboratory media; however, the monophagous species described by Pipkin *et al.* with one exception refused to oviposit, or underwent a prolonged larval and pupal state and failed to emerge as adults. In other words, the monophagous species is usually entirely adapted to its own specialized niche, whereas polyphagous species have less specialized requirements.

Similarly, until recently it was almost impossible to culture most Hawaiian species, which are presumably highly specialized. A food medium has, however, been devised enabling some species to be reared (see Carson *et al.* 1970). Nevertheless, until it is possible to rear many more Hawaiian species in the laboratory, many aspects of the over-all evolutionary biology of the Hawaiian drosophiloids must necessarily remain obscure. As is clear from a perusal of the previous chapters of this book, much of the information presented is dependent on ease of breeding in the laboratory. Some work has been carried out on possible media for the Hawaiian species; for example, Robertson, Shook, Takei, and Gaines (1968) fractionated the *Cheirodendron* leaf into three parts—(1) combined organic solvents, (2) water-soluble constituents, and (3) principally cellulose. Adding (1) to agar induced some females of *D. disticha* to oviposit, which they refused to do with plain agar. Larvae were reared to the second instar by various combinations, including yeasts found in the crops of adults, but never to maturity. It was concluded that *D. disticha* differs rather extensively from the species of *Drosophila* so far examined from the metabolic point of view. Kircher (see Carson *et al.* 1970) has advanced the view that the leaf-breeders in Hawaii may have evolved to utilize the slow release of nutrients caused by the bacterial decay of *Cheirodendron* leaves.

Some recent work (Kambysellis and Heed 1971) shows the dependence of the precise larval niches in determining the fecundity potential of the Hawaiian Drosophilidae. The fecundity potential was analysed from the number of ovarioles per fly and the number of mature eggs per ovariole. In a niche where flies feed on flowers, with the consequent limited food supply, the number of ovarioles is minimized, and their function is modified to alternating ovariole development so that at most one mature egg is deposited daily. Further, the restricted life span of the flower as a nutritional source has been compensated by a prolonged retention of the egg in the vagina, so that advanced stages of embryonic development are reached before oviposition. This is followed by a brief larval stage so that the feeding time to full larval growth is minimized. In a second niche (the leaf niche) the supply of nutrition, being primarily bacteria, is higher than in the flower niche, partly because the number of leaves per unit ground is higher than for flowers, partly because their suitability as a nutritional source is prolonged, and partly because the rather constant temperature and humidity of these habitats provide suitable conditions for oviposition throughout the year. The response to these conditions is a stable, high population-density throughout the year, which has been maintained by a relatively low fecundity potential for the species, which has evolved by the development of asynchronous ovariole development, a relatively low number of ovarioles, and the restriction of the number of eggs per ovariole to one. In a third niche, rich in larval nutrition, such as is available to those breeding on decaying leaf stems with yeast, the reproductive potential may be increased by increasing the number of ovarioles per fly, developing all ovarioles simultaneously, and maintaining one mature egg per ovariole. However, when the breeding sites are nutritionally rich, yet infrequent so that the flies have to search for them, the species respond by increasing their number of ovarioles and by accumulating numerous small eggs in each ovariole. When a food source is located, numerous eggs are deposited, usually in clusters.

D. buzzatii is of interest as it is a widespread species with niche specificity. It originated in South America in association with one or more species of the cactus genus *Opuntia*. This cactus, itself of South American origin, has become a weed in various parts of the world, and *D. buzzatii* has apparently gone along with it. There is no evidence that the species is able to breed outside its cactus niche (Carson and Wasserman 1965). It is of interest that throughout its spread it has carried a particular inversion on chromosome II, namely IIj. In Argentina the population shows polymorphism on chromosome II for two inversions in addition to IIj. This is the presumed site of endemic origin, and during its spread it took a reduced level of chromosomal polymorphism —that is, its genetic structure then corresponded more to a marginal than a central population (see §12.2). Another species, *D. pachea*, is restricted to a species of cactus, *Lophocereus schottii*, which breeds throughout the Sonoran Desert of Mexico. The reason seems to be that it requires the cactus as a dietary supplement. A unique sterol, Δ^7-stigmasten-3β-ol (schottenol), either isolated from the cactus or synthesized, can replace the cactus in the diet of the flies. Sterols are of importance in the diet of insects, being probable precursors of the moulting hormone ecdysone, and may be involved in female fertility. Normally cholesterol satisfies this requirement, but *D. pachea* cannot use this sterol (Heed and Kircher 1965). Later work showed that the alkaloid pilocereine, derived from the cactus, and the cactus itself, killed the adults and/or progeny of eight other species but not *D. pachea*. The ability of *D. pachea* to tolerate pilocereine and its absolute requirement for schottenol probably explains the unique cactus–*Drosophila* relationship in the Sonoran desert (Kircher, Heed, Russell, and Grove 1967).

So far, our discussion of niche specificity mainly involves plants; however, associations have been found between *Drosophila carcinophila* and the land crab *Gecarcinus ruricola*, such that the fly breeds as a commensal on the exterior nephric nerves of the crab (Carson 1967). Another species, *D. endobranchia*, which is only moderately closely related to *D. carcinophila*, has been collected from *G. ruricola* and *G. lateralis* (Carson and Wheeler 1967).

For most species of *Drosophila* our knowledge is much more restricted, and this tends to apply to some of the well-studied laboratory species—for example, the sibling species *D. pseudoobscura* and *D. persimilis*, which in nature have been found on the infected sap exudates (slime fluxes) of *Quercus kelloggii*, the Californian black oak (Carson 1951). Occasionally *D. victoria* and/or *D. californica* were also found breeding in the same flux. A study of the organisms frequenting slime fluxes led to the working hypothesis that of the slime flux biota, predators and associates are the main biotic factors preventing excessive oviposition by *Drosophila* at the slime flux. As a result, it is probable that competitive relationships between the various organisms associated with *Drosophila* in the exploitation of the slime flux seldom occur (Sokoloff 1964). Other isolated observations consist of the finding of pupae of the *montana* complex in the rotting phloem of aspen logs (Spieth 1951), and the finding of larvae of the *americana* complex in the stumps of *Salix interior* (Blight and Romano 1953). Another commonly studied species, *D. robusta*, has a wide latitude of breeding sites, but there appears to be a distinct tendency for the American elm (*Ulmus americana*) to serve as the principal host tree, and in fact the distribution of *D. robusta* parallels *U. americana*. There are some reports of the elm fluxes as breeding sites for *D. robusta* (see Carson 1958a).

Information on the cosmopolitan species *D. melanogaster* and its sibling species *D. simulans* is scanty, but they congregate and breed in overripe fruit. Hoenigsberg (1968) described a situation in one site in Colombia where *D. melanogaster* was ten times more frequent than *D. simulans* in 1963, while 5 years later in 1967 *D. simulans* had become rather more common than *D. melanogaster*. Some ecological changes occurred in the vegetation due to the construction of a road, but whether this was the mechanism which could account for the change is unknown. A similar observation was made by

Tantawy, Mourad, and Masri (1970), who found that *D. melanogaster* lost ground in competition with *D. simulans* over the period 1956–66 in the area around Alexandria, Egypt. It is known that *D. melanogaster* tends to be more common than *D. simulans* in the northern U.S.A., and while in the south the reverse occurs (Wallace 1968). In Brazil, Freire-Maia (1964) reported that *D. simulans* is commoner in the colder regions than *D. melanogaster*, which differs from Wallace's (1968) observation. The explanation may partly lie in the degree of temperature fluctuations that the two species can tolerate in the wild (§8.1). McKenzie and Parsons (1972) have recently found that adult *D. melanogaster* are tolerant of exposure to food containing 9 per cent ethanol for six days, whereas this is lethal for *D. simulans*; and further, although less spectacular a finding, that the larval-to-adult mortality is lower in the former species. Preliminary evidence from Chateau Tahbilk, a winery 60 miles from Melbourne, shows that outside the winery both species occur, but that inside *D. melanogaster* occurs almost exclusively.

2. Niche breadth

It is not the function of this book to enter into detailed discussions of the term niche, as this has been examined on a number of occasions in recent years (see, for example, MacArthur 1968). Even so, although the term has already been used in the book, and in particular in the previous section, some clarification seems desirable at this stage so that the concept of niche breadth can be examined. The term arose out of the consideration by some writers of the role of the animal in the community and, by other writers, of the subdivisions of the environment. In Hutchinson's (1957) niche, each measurable feature of the environment was given one co-ordinate in an infinite-dimensional space, and the region in which the fitness of an organism was positive was called that individual's niche. Even though this definition is, in absolute terms, difficult to assess, it does permit statements to be made about *differences between niches* and, as is shown later, some quantification of these differences can be achieved. The niches of two similar species can be compared, as can the niche of one species at two places or at two times.

MacArthur (1968) pointed out that there is a parallel between the term niche and phenotype. The phenotype represents all the measurements that can be made on an individual during its lifetime, whether they be morphological, behavioural, direct fitness traits such as viability, or those measurements which constitute the niche. Again, differences between phenotypes are easily assessable, although defining phenotype in absolute terms is difficult once one departs from purely morphological criteria.

The concept of the niche thus includes all the relationships of an organism to the physical and biotic environment. So defined, it includes the physical conditions of light, temperature, humidity, etc., required for the survival of the organism, as well as its requirement for food, a place to live, and its relationships with other organisms. In this sense, the term is most frequently used in the biological literature (Ayala 1970a). It is clear that the distribution of the species of *Drosophila* that have been under discussion depends on all the above factors, as well as many others. There are, however, difficulties when attempting to apply a term such as the niche to the terrestrial environment in particular. Thus, some species may exist fairly uniformly over an environment and so therefore show considerable niche breadth. On the other hand, a second species may spend most of its time on one resource or component of the environment, and neglect other resources or components; this second species would therefore show less niche breadth than the former. Examples of both types of species are described in the previous section. Levins (1968) mentioned a number of approaches to the question of assessing niche breadth in different *Drosophila* species.

One approach is that of Tantawy and Mallah (1961) where it was shown that *D. simulans* has a narrower, more specialized temperature niche than

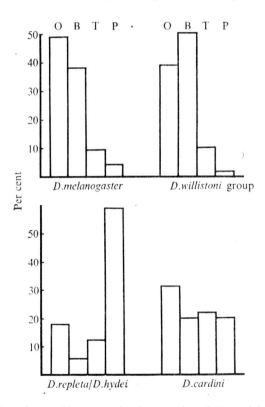

FIG. 14.1. Food preference histograms for four species of *Drosophila*: O, Orange, B, banana; T, tomato; P, potato. (After Levins 1968.)

D. melanogaster (§10.1). Any measure in this sense can be used as a quantification of niche breadth. A second approach consists of habitat or food selectivity in multiple environments, thus several kinds of bait (banana, tomato, potato, and orange) were set up in *Drosophila* traps less than ten feet apart. Some results are given in Fig. 14.1 A species such as *D. cardini*, which is attracted to each bait with almost equal frequency, would be said to have a broad niche for food as compared with *D. repleta* (for those foods chosen). The third approach consists of looking at the actual distribution of species over a number of traps in a small area. From this, estimates of niche breadth can be made. Such an approach, if carried out systematically at different times and localities, might well provide detailed information of a microecological type on the distribution of various cosmopolitan species. Uniformity of distribution over a patchy environment indicates a broad niche, and extreme clustering suggests a narrow niche. However, a clustering tendency *per se*, or the persistence of progeny at the site of their birth, which is a real problem in many *Drosophila* species judging by their rather low migration rates (§10.1) may affect results. Assuming then, that the environment sampled is rather small for the mobility of the species, the niche breadth can be measured.

A useful method of measurement of niche breadth comprises a procedure used initially when examining the abundance of species on the basis of information theory, and we shall first develop the argument in terms of the relative abundance of species (MacArthur 1965; see also Pielou 1969). The question asked is that of the difficulty of predicting correctly the species of the next individual to be collected at a particular site. Communication engineers face this problem, since they are interested in the difficulty of predicting correctly the identity of the next letter in a message. If successive letters are chosen independently, the formula

$$H = -\sum_{i=1}^{N} p_i \log_e p_i$$

measures the uncertainty of the next letter, where p_i is the probability of the *i*th letter of the alphabet. Similarly, if successive individuals in a census are independent of previous ones, the same formula is applicable where N is the number of species in the count and p_i is the proportion of the total number of individuals which belong to the *i*th species (MacArthur 1965). In this calculation $\sum_{i=1}^{N} p_i = 1$. Thus, if all N species are equally common, then each is a proportion $\frac{1}{N}$ of the total, and H becomes $-N\left(\frac{1}{N}\log_e \frac{1}{N}\right) = \log_e N$. If we take an example of a census of 99 individuals of one species and 1 of a second, we calculate:

$$H = -p_1 \log_e p_1 - p_2 \log_e p_2 = 0.056,$$

and for a census of 50 individuals of each of the two species, $H=\log_e 2=0\cdot693$. If we take $e^{0\cdot056}=1\cdot057$ for the first census and $e^{0\cdot693}=2$ for the second, we see that for a situation of N equally common specie

$$E=\exp-\sum_{i=1}^{N} p_i \log_e p_i = \exp H$$

gives a value of N, and if the species are not equally common, this expression will be arithmetically less than N. Thus both H and E provide measures of species diversity, regarding a situation of 50 individuals of each species as more diverse than where there is 99 of one species and 1 individual of the other. It should finally be pointed out that this method is only applicable for a population large enough to be effectively infinite, or at least too large for its members to be counted (Pielou 1966). This is presumably a reasonable approximation for the more cosmopolitan species of *Drosophila*.

Now, this argument can be used for measures of niche breadth from a knowledge of the abundances of the species in components of the habitat. Thus the niche breadth B of a species can be expressed as formulae analogous to H and E, but in which p_i represents the proportion of the given species in environment i (Levins 1968). In an analogous way, measures of community diversity can also be computed.

Using a measure of community diversity, Levins found for his own and other published data that there is a relatively small contribution due to geographic site in *D. melanogaster*, which he suggests is a result peculiar to small animals. Within a forest in Puerto Rico, for example, there is considerable variation in temperature and humidity over very short distances. Thus it is likely that almost all the range of environments that can be found among forests at all elevations in Puerto Rico can occur somewhere within a limited region in the same forest. Similarly, the diversity found at Austin, Texas is only increased by one-third when the whole south-west of the United States is included. This would not be true for species with very narrow food preferences, some of which are mentioned in §14.1.

In laboratory experiments in several species of *Drosophila*, it has been found that when offered a series of apparently similar oviposition sites flies do not distribute their eggs at random. The main factor bringing about the aggregation of the eggs is the preference of female flies for ovipositioning near the sites previously used for oviposition by other females. It is further possible to direct the flies to a given site simply by placing some eggs on one of the several suitable sites offered to them (see del Solar and Palomino 1966, del Solar 1968). Mainardi (1968) showed that single inseminated females preferred to lay eggs where males had previously been present as compared with sites where males had not previously been present, when given a choice of the two types of sites. This indicates that female flies can recognize the previous presence of males in a site. In *D. pseudoobscura*, the degree of aggregation can be changed by selection and so is genetically controlled. Probably there is a compromise between a single oviposition site, leaving all other potential food sources unexploited, and the situation of having many sites with so few eggs deposited that the food is not properly conditioned by the larvae. Natural selection strikes the compromise between some oviposition sites receiving larger numbers of eggs than others, and others remaining unoccupied. This mechanism may well also operate in the wild and be an important controlling factor, but at the moment this must remain pure supposition. Moore (1952a) has presented evidence to show that in competition experiments between the sibling species *D. simulans* and *D. melanogaster*, *D. simulans* is better able to oviposit in the centre of food cups and on food having a surface crust; in other words, desiccation makes the medium less favourable for *D. melanogaster* as compared with *D. simulans* (see also Sameoto and Miller 1966). It turned out to be possible to reverse the outcome of competition by taking advantage of oviposition site preferences. Extrapolating to the wild, it seems reasonable that sibling species should have somewhat different preferences for oviposition sites. Other data have shown variations in pupation site in *D. melanogaster* (Wallace 1968) for different strains, and also the proportion of drowned pupae in a medium which tends to liquify was found to vary between strains. Variations in pupation site under genetic control have also been reported in *D. willistoni* (§12.1). Predictably, pupation sites can be varied environmentally. Some factors are variations in larval-competition levels, food moisture-content, and temperature (Sokal, Ehrlich, Hunter, and Schlager 1960). This shows the need to control such variables, otherwise genotype × environment interactions could complicate conclusions. These traits may be relevant in determining survival in natural conditions, and again one would expect variations between sibling and other closely related species, so showing variations in niches occupied.

3. Interspecific competitive ability

A number of laboratory experiments, the relevance of which is rather difficult to assess from the ecological point of view in the wild, have been carried out on interspecific competitive ability in *Drosophila*. Thus *D. funebris* and *D. melanogaster* can coexist at 20°C for long periods of time. Merrell (1951) found that *D. melanogaster* is favoured by the addition of fresh food, while the proportion of *D. funebris* increases with the age of the food. Adding food at regular intervals leads to a relatively stable equilibrium with *D. melanogaster* predominating. The two species coexisted until the experiment was terminated after nearly two years of 30 generations. Although the two species differ ecologically, there is no doubt that they competed for the same basic source of energy. For the species *D. nebulosa* and *D. serrata*, it was found that both species coexisted at 19°C with *D. nebulosa* predominating, but at 25°C *D. serrata* was eliminated in a few generations (Ayala 1966c).

For the sibling species *D. melanogaster* and *D. simulans*, Moore (1952a, b) found in population cages at 25°C that *D. simulans* was eliminated in nineteen of twenty cages in about 100 days. This did not occur in the twentieth cage where *D. simulans* remained for considerably longer with improved competitive ability developed by selection. Futuyma (1970) found a variety of responses using different populations of *D. melanogaster* with an inbred strain of *D. simulans*, the explanation of which requires invoking qualitatively different changes in different populations. Futuyma concluded that the genetic variance for traits affecting the outcome of competition is mainly non-additive, so preventing rapid adaptation to environmental change. For the same pair of species, Barker and Podger (1970) found that *D. melanogaster* raised in mixed species cultures were less fecund than those from pure cultures, while *D. simulans* showed the reverse effect. Therefore, from these results it seems that predictions of what may happen in competition may be difficult to make from performance in pure culture. The results of Futuyma would seem to indicate the likelihood of different results with different strains *within* species, because of genetic differences between strains. Because genetic differences occur, it is reasonable to suppose that competitive ability might change with selection.

Tantawy and El-Wakil (1970) studied competition between *D. funebris* and *D. virilis* as well as fitnesses of populations of the two species in isolation both collected from the University Farm, Alexandria, Egypt. On various fitness parameters, namely egg production, percentage hatchability, and longevity, *D. funebris* was superior to *D. virilis* in isolation. However, in a competition experiment in population cages, *D. funebris* was rapidly eliminated by *D. virilis* in 90 days, in spite of the superior fitness of the former in isolation. This situation cannot be explained on the data so far presented, but again shows that results in pure culture are not necessarily predictors of results in competition.

In contrast, however, are data (Table 14.1) of Ayala (1970b) who studied population sizes, as a measure of adaptedness, of *D. serrata* from three geographic localities, Popondetta, Cooktown, and Sydney, by their performance in interspecific competition with three other species of *Drosophila*—namely, *D. pseudoobscura*, *D. melanogaster*, and *D. nebulosa*. The experiments with *D. pseudoobscura* and *D. melanogaster* were carried out at 25°C, and that with *D. nebulosa* at 19°C. In the two experiments at 25°C the Cooktown strain of *D. serrata* had the highest fitness, while the other two strains, Popondetta and Sydney, did not differ significantly from each other. In competition at 19°C, the Cooktown strain had the highest fitness, Sydney being intermediate, and Popondetta the lowest. The adaptednesses of these three strains of *D. serrata* have been studied in single-species populations (§12.3) giving results consistent with these. That is, in single-species populations, the Cooktown strain had the highest fitness at 25°C and 19°C, and the fitness of Popondetta and Sydney did not differ significantly at 25°C, but at 19°C Sydney had a higher fitness than Popondetta. Similar experiments have been carried out comparing the performance of polymorphic and monomorphic populations of *D. pseudoobscura* when they compete with *D. serrata*. The *D. serrata* strain was from Popondetta, and *D. pseudoobscura* flies were either monomorphic for the *CH* or the *AR* chromosomes, or polymorphic for *CH* and *AR*. At 23·5°C and 25°C, it was found (Ayala 1968a, 1969b) that the adaptedness of *D. pseudoobscura* was greater in the monomorphic *AR* than in the monomorphic *CH* population, and greater in the polymorphic than in either monomorphic population. This corresponds with the sequence of fitnesses

TABLE 14.1

Fitnesses (W) of the strains of D. serrata *relative to the Popondetta strain as measured by competition with strains of* D. pseudoobscura, D. melanogaster, *and* D. nebulosa

	D. pseudoobscura (25°C)	D. melanogaster (25°C)	D. nebulosa (19°C)
Popondetta (New Guinea)	1	1	1
Cooktown (Queensland)	1·408†	1·109‡	1·216†
Sydney (New South Wales)	0·965	1·026	1·102‡

† $P<0.001$ and ‡ $P<0.05$ for deviation of W from unity.

After Ayala 1970b

reported at 25°C in population cage experiments, and data on the innate capacity for increase (§11.2) with just the species *D. pseudoobscura* alone (Dobzhansky 1948a). Thus once again there is a correspondence between the adaptednesses obtained from single species populations with those based on interspecific phenomena. Clearly further investigation is needed, in view of the fact that other published results do not support this. It must also be pointed out that the precise technical details of maintaining long-term competition experiments could well lead to variable results.

Like Moore (1952b) in studies of competition between *D. melanogaster* and *D. simulans*, Ayala (1969c) in studies of competition between *D. serrata* and *D. nebulosa* found that the interspecific competitive abilities of both species improved over a number of generations. This seems to indicate that when two species compete for certain limited resources, natural selection may produce genotypes which are better competitors for that resource. It is also possible that selection could be towards the avoidance of competition—that is, for genotypes which allow one species to exploit resources not utilized by the competing species. Selection for avoidance of competition presumably increases the probability of coexistence, and leads to ecological divergence of two coexisting species. This implies that it is likely that sibling species in particular might well reveal ecological differences, if studied intensively enough; this is discussed in further detail in the next chapter. It would seem likely that both species may share some of the available resources of food and space, while each species may exploit resources that the other species cannot use. In the laboratory, *Drosophila* populations of two species may share some food resources, such as carbohydrates and certain yeasts species, and also the same available space. But, from the considerations already presented, it is likely that the larvae of one species may eat some yeast or mould species not exploited by the other, and vice versa. Similarly the adults may use different components of available space and show oviposition site variations; the same might be true for pupation sites. These differences might be expected to be more marked under natural than laboratory environments.

The notion that two species with similar ecological requirements cannot coexist indefinitely was formulated early in this century. Mathematical investigations led to the conclusion that two species cannot coexist at an equilibrium if they utilize a common resource available in a limited amount. This result has become known as the Principle of Competitive Exclusion, or Gause's Principle, so named after Gause (1934) who reviewed the mathematical evidence and provided experimental and field observations supporting the conclusions derived from them. Thus it can be said from this principle that two species with identical niches cannot coexist. Or, conversely, two coexisting species must occupy different niches. Clearly, assuming the niche to be defined in the broadest sense, as including all the ecological relationships of the organism, it is true that no two species will occupy the same niche, and that if two species persist together indefinitely in a steady state, then there must be some ecological distinction between them.

As we have seen in this book and especially in this chapter, many related species of *Drosophila* coexist in nature. Clearly, it is impossible to say that there are no ecological differences between coexisting species in nature, since natural environments are far too complex to make such an assessment. In the laboratory, however, some evidence has been presented for coexistence between species, and furthermore the ability to coexist seems to be subject to natural selection.

Ayala (1969*d*) discussed the issue in more detail in various populations of *D. pseudoobscura* and *D. serrata*. At 25°C *D. serrata* excludes *D. pseudoobscura* in a few generations, while at 19°C *D. serrata* is rapidly eliminated (Ayala 1969*d*). On the other hand, at 23·5°C the two species can coexist, apparently at an equilibrium at frequencies dependent on the genetic composition of the particular strains involved within species. The dependence of the result on temperature is not surprising in view of the extreme sensitivity of many fitness parameters to temperature. Table 14.2 gives the mean productivity and mean population sizes of six AR and five CH populations both with *D. serrata* at 23·5°C, from which it can be seen that the frequency of AR relative to *D. serrata* is higher than that of CH. This agress with the interspecific and competition data referred to above showing AR to have a greater adaptedness than CH. From these results, and a consideration of the Volterra (1926) equations, which describe competition between two

TABLE 14.2

Mean number of flies of D. serrata *and* D. pseudoobscura *in the 'competition' populations*

Population	Species	Productivity	Population size
44	*D. serrata*	125±14	262±25
	D. pseudoobscura AR	105±10	255±24
45	*D. serrata*	101±8	187±11
	D. pseudoobscura AR	119±8	288±17
46	*D. serrata*	162±15	331±27
	D. pseudoobscura AR	92±9	207±20
47	*D. serrata*	131±8	292±14
	D. pseudoobscura AR	108±8	279±18
48	*D. serrata*	129±9	289±18
	D. pseudoobscura AR	107±8	261±17
49	*D. serrata*	149±9	304±13
	D. pseudoobscura AR	97±8	223±16
Average	*D. serrata*	133·0±7·5	277·6±12·8
	D. pseudoobscura AR	104·6±7·1	252·0±16·8
64	*D. serrata*	268±15	547±27
	D. pseudoobscura CH	61±8	129±15
66	*D. serrata*	245±18	539±33
	D. pseudoobscura CH	78±10	145±17
67	*D. serrata*	269±20	588±33
	D. pseudoobscura CH	74±7	149±14
68	*D. serrata*	248±19	562±38
	D. pseudoobscura CH	55±7	125±13
69	*D. serrata*	254±19	508±36
	D. pseudoobscura CH	49±12	110±20
Average	*D. serrata*	257·0±13·5	549·1±25·6
	D. pseudoobscura CH	63·3±6·2	131·7±13·1

Size of population was measured after addition of the young flies.

After Ayala 1969*d*

species in the same microcosm, Ayala (1969d) was able to conclude that in fact two species competing for limited resources may coexist, which he considers invalidates the Principle of Competitive Exclusion. This conclusion has already provoked some discussion in the literature (e.g., Borowsky 1971), as might be expected, and more can be anticipated. Some of the biological mechanisms Ayala (1970a) put forward leading to competitive coexistence are: frequency-dependent fitnesses such that species fitnesses depend on their relative numbers; fitness interactions between stages of the life cycle, such that one species may be a better competitor at one stage of the life cycle and worse at another; and oscillations in competitive ability from one generation to the next, such as seasonal changes (in nature).

The main conclusion seems to be that the Principle of Competitive Exclusion, as stated in its simplest form, may in fact be an oversimplification. Much more work is necessary to find out the conditions under which closely related species are able to coexist, and some of the points above put forward by Ayala are in need of much more study. Looked at in another way, if two species coexist, then there must be some, even if very subtle, differences between them. In this sense, the Principle of Competitive Exclusion seems trivial, since to *prove* no distinction seems impossible. The other point is that competitive ability is not static, but evolves, by improving the ability of one species to exploit the resources of a second species, or alternatively by selecting for the avoidance of competition whereby the two species exploit slightly different niches. Further work will be awaited with interest on the conditions under which species can coexist, and the evolutionary implications of coexistence. Whether Gause's principle is considered to be proven or disproven by this work seems less important than the general study of interrelationships in artificial and natural environments of closely related species.

15
BEHAVIOURAL AND ECOLOGICAL ISOLATION

THE mechanisms isolating closely related species from each other—that is, preventing gene exchange between them—depend on a complex of behavioural and ecological factors when the species occur sympatrically. For example, *D. pseudoobscura* and *D. persimilis* form a pair of sibling species, sympatric in some regions. Where the species are sympatric, isolation is maintained by several factors, some of which are as follows:

(1) somewhat different habitat preferences such that *D. persimilis* occurs in cooler niches and *D. pseudoobscura* in warmer, drier niches;

(2) some differences in food preferences;

(3) different diurnal periods of maximum activity: thus Dobzhansky *et al.* (1956) recorded for the Yosemite region of California that among the flies attracted to the yeasted baits in the morning the proportion of *D. pseudoobscura* was lower, and that of *D. persimilis* higher, than among flies attracted during the evening period of activity, while the reverse was found for the evening (Table 15.1);

TABLE 15.1

Numbers of flies collected in the morning and evening in the Yosemite region of California

Month	Morning		Evening	
	pseudoobscura	*persimilis*	*pseudoobscura*	*persimilis*
June	68	111	682	432
July	210	297	694	446
August	65	75	681	443

After Dobzhansky *et al.* 1956

(4) sexual isolation, associated with which are male courtship songs which vary between the two species (Ewing 1969). Thus males of *D. pseudoobscura* produce two songs controlled by the wings, a low repetition-rate song which consists of trains of 525 Hz pulses at six per second, and a high repetition-rate song in which 250 Hz pulses are repeated at twenty-four times per second. In *D. persimilis* the low repetition-rate song is absent or occurs in a very

abbreviated form, and the high repetition-rate song consists of 525 Hz pulses repeated at fifteen times per second. By carrying out F_1 and back-crosses between the species, it was found that the presence of the low repetition-rate song, and the frequency within pulse of the high repetition-rate songs are controlled by genes on the X chromosome. The pulse repetition-rate of the high repetition-rate song was found to be inherited independently of the other traits.

The first three factors promoting isolation, which are ecological-physiological but obviously with behavioural components, are not entirely effective as both species have been found feeding side by side in the same slime flux on the black oak *Quercus kellogii* (Carson 1951), so that the absence of interspecific matings in natural habitats is perhaps largely due to ethological isolation (Dobzhansky 1951). Furthermore, when interspecific matings take place in the laboratory, fewer sperm are transferred than in intraspecific crosses, the F_1 males are sterile, and the F_1 female progeny from back-crosses have reduced vigour.

In the laboratory, hybridization will occur relatively readily, but virgins are usually about four days old when utilized in many experiments. However, when flies of both sexes and species were placed together a few hours after emergence the proportion of hybrids was lower. Spieth (1958) considers that this higher level of isolation may be due to individuals of both species maturing together, and acquiring the ability to discriminate between species before becoming sexually mature. In further experiments, it was found that a *D. persimilis* female, once having mated with a *D. persimilis* male, will not accept a *D. pseudoobscura* male subsequently. Thus the high level of isolation found between these two species is due to ethological isolation enhanced by various other pre-mating isolating mechanisms as well as by hybrid male sterility, a post-mating isolating mechanism. Under laboratory conditions, which differ from those prevailing in the wild, the degree of isolation may be somewhat reduced. Finally, it seems likely that isolation in the wild may not only be innate but partially learned.

Laboratory experiments also indicate some other variables that may be relevant. Thus, Mayr and Dobzhansky (1945) found that the degree of isolation was temperature-dependent, and was relatively low for flies raised at 16·5°C. Whether this has any association with male courtship song variations is difficult to say, but pulse repetition-rate is temperature-dependent in *D. melanogaster*, where it has been found to increase by 1·4 pulses for each rise of 1°C (Shorey 1962). Although isolation is relatively low in the laboratory between *D. pseudoobscura* and *D. persimilis* at 16·5°C, it can be rapidly increased by selection. Thus Koopman (1950) set up artificial mixed populations of the two species, using marker stocks, and carried out selection in each generation for pure-bred flies, that is, the progeny of parents that had been homogamically mated. Isolation developed rapidly after a few generations, and part of the reproductive isolation that developed was shown by sexual isolation tests to be behavioural. The experiments also show that even though the level of sexual isolation depends on the environment this can be altered remarkably quickly by selection. Koopman aided the change by removing the hybrids each generation, in this way simulating complete hybrid inviability.

Later work (Kessler 1966) confirmed and extended some of Koopman's findings. He selected artificially for weaker and for stronger ethological isolation between the same pair of species. To weaken the isolation, he selected females and males that mated soonest with individuals of the opposite sex of the other species, and to strengthen the isolation individuals were selected which failed to mate with the other species and then mated rapidly within species. After eighteen generations of selection, significant increases and decreases of the levels of isolation were obtained. In Table 15.2 the results of multiple-choice experiments are given with joint isolation indices, for the various unrelated flies and those selected for high (H) and low (L) degrees of isolation. It will be noted that $H \times H$ matings show more isolation than the unselecteds and $L \times L$ show less, while the $H \times L$ and $L \times H$ are close to the unselected especially $H♀ \times L♂$. Thus ethological isolation has increased in the high lines and decreased in the low lines. Divergence was found in both sexes for both species, except for the low-isolation females in *D. persimilis*.

Turning to *D. melanogaster* and *D. simulans*, a further pair of sibling species, it is likely that isolation is due to the following factors: (1) *D. melanogaster* can probably tolerate a greater range of temperatures than *D. simulans* (§8.1); (2) *D. melanogaster* is more tolerant of ethanol, especially as adults (§14.1); (3) some oviposition site variations occur in laboratory experiments (§14.2); and (4) sexual isolation, as shown by tests in the laboratory (Sturtevant 1929, Barker 1962*b*). Ethological isolating mechanisms almost completely prevent matings, and such hybrids that are formed are sterile. Where matings occur in the laboratory, genotype and age are important. In mass cultures, more interspecific matings occur than when the flies are set up as single pairs, perhaps because of facilitation among courting males whereby one courtship stimulates the other males in the same culture to increased activity. Later data (Barker 1967) showed that the frequency of interspecific hybridization increased with increase in the proportion of males. Bennet-Clark and Ewing (1969) analysed the pulse interval of these two species (Fig. 15.1), and found that for *D. simulans* to be longer than that for *D. melanogaster*. Earlier work (Manning 1959) showed that the sexual behaviour of the two types of males can be readily classified into the same basic elements (see §3.1), but that males of *D. simulans* have longer lag periods before courtship, and longer bouts of simple orientation; in other words, the courtship behaviour of *D. melanogaster* is more active than that of *D. simulans*. The wing display of *D. simulans* is mainly scissoring, and that of *D. melanogaster*

TABLE 15.2

Frequency of mating in multiple-choice tests in selected and control lines in D. pseudoobscura *and* D. persimilis

Degree of isolation		Type of mating (per cent)				Joint isolation index
Females	Males	pseudoobscura ♀ pseudoobscura ♂	pseudoobscura ♀ persimilis ♂	persimilis ♀ pseudoobscura ♂	persimilis ♀ persimilis ♂	
Unselected	Unselected	49·0	5·9	3·9	41·2	0·804 ± 0·059
Low	Low	26·8	20·7	1·4	51·2	0·559 ± 0·057
Low	High	26·3	14·9	0·0	58·9	0·703 ± 0·054
High	Low	40·1	7·6	1·2	51·2	0·826 ± 0·043
High	High	39·2	2·0	0·0	58·8	0·961 ± 0·027

After Kessler 1966

is vibration: it seems that scissoring represents a lower level of sexual excitation than vibration, since with lower stimulation *D. melanogaster* males show an increased proportion of scissoring and with increased stimulation *D. simulans* males show more vibration. There is therefore no difference in the basic organization of the sexual behaviour of the two types of males. The difference that exists is quantitative in that *D. simulans* males have a slower rise to sexual excitation than *D. melanogaster*. So far as females are concerned, the females of *D. simulans* are more responsive to the visual aspects of the male's courtship, and less responsive to those stimuli perceived by their antennae, than are *D. melanogaster* females.

Two South American sibling species, *D. gaucha* and *D. pavani* (Koref-Santibañez 1964), show little discrimination in courtship behaviour, and

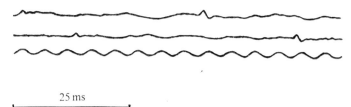

Fig. 15.1. Oscillograms of tape recordings of courtship songs of *D. melanogaster* (top trace) and *D. simulans* (middle trace) at 25°C. The bottom trace is a 200-Hz time marker. The pulse interval, measured from the beginning of one pulse to the beginning of the next, is 34 msec in *D. melanogaster* and 48 msec in *D. simulans*. (After Bennet-Clark and Ewing 1969.)

crosses yield abundant but sterile progenies. Thus the main isolating mechanism is hybrid sterility. In nature the lack of discrimination in courtship behaviour is unimportant, since except for a small part of Argentina where the two species are found together they are allopatric. It may be expected that greater divergence in courtship behaviour would occur in sympatric rather than allopatric species (see, for example, the evidence in §7.3 for allopatric and sympatric pairs of races in *D. paulistorum*) so as to avoid the production of large numbers of relatively unfit hybrids.

Dobzhansky, Ehrman, and Kastritsis (1968) found the closely related sympatric American species, *D. pseudoobscura*, *D. persimilis*, and *D. miranda*, to show strong ethological isolation, as do the closely related sympatric Japanese species, *D. bifasciata* and *D. imaii*. The ethological isolation between the sympatric American and the sympatric Japanese species was at least as strong as between the allopatric American and Japanese species, and between either of these and the European *D. subobscura*. Thus, the degree of ethological isolation is at least as strong between closely related sympatric species as between the less closely related allopatric species; in other words, in sympatric species ethological isolation is enhanced by natural selection, so preventing the wastage of reproductive potential.

These various examples and others show some of the mechanisms preventing gene exchange, which are made up of a complex of behavioural and ecological factors. The strongest and most obvious selective forces occur in closely related sympatric species, where ethological isolation is selected very strongly. Quite an amount of work remains to be done on these topics, to unravel in more detail the complexity of factors controlling isolation between closely related species in the wild, but the combined approach of laboratory experimentation and field observation should be productive.

The differences in courtship between closely related species are generally quantitative, but larger than the differences between mutants within species which tend to be very small indeed. Thus, in *D. melanogaster* the behavioural differences between wild-type and yellow (Bastock 1956), and between ebony and vestigial compared with wild-type (Crossley 1964,† Crossley and Zuill 1970), are very small compared with the differences between sibling species— for example, *D. melanogaster* and *D. simulans*. Brown (1965) quantified differences between eleven species of the *obscura* group for behavioural and morphological characters, by working out 'Mean Character Differences' between pairs of species based on twenty behavioural and twenty-four morphological traits. Although there is a certain amount of arbitrariness in the calculation of the Mean Character Differences, a correlation coefficient between the fifty-five possible species combinations in pairs between behavioural and morphological traits came to 0·576 ($P<0·001$ for difference from 0), showing quite clearly that behavioural and morphological divergence are associated. Brown's (1965) summary of the situation is: 'It is clear that, throughout the genus *Drosophila*, there is a close correlation between behavioural and morphological divergence. The differences between mutants are slight; those between sibling species are greater, and those between non-sibling species in the same division greater still. As the morphological divergence increases between divisions, and between groups of the same subgenus, so does that of the behaviour, until in the end we find the major differences in behaviour between the subgenera.'

Spieth (1968) and Carson *et al.* (1970) have reported on the courtship behaviour of the Hawaiian species, which have a number of unique features, but even so it conforms in some ways to the general pattern of drosophiloids from other parts of the world. In brief, the usual world-wide pattern involves a diurnal morning and evening assembling on the feeding sites by adults of all ages. The females devote most of their time to feeding, but the males after feeding for a short time then move actively about, investigating other individuals of about their own size and facies. Thus, the males will orientate and initiate at least the beginning of courtship with any male or female of their own species or of other similar-appearing species. If a female of his

† Crossley, S. (1964). *An experimental study of sexual isolation within a species of* Drosophila. (D.Phil. thesis, University of Oxford)

own species has been approached, the male will then proceed to court in typical fashion. At any given feeding site, most females will either be too young to accept the males' overtures, or else have been previously fertilized. Of the numerous courtships attempted by males, only a small percentage results in copulation, and after courtship display males of most non-Hawaiian species try to mount non-receptive females. There is therefore much activity at the feeding site, due to the persistent courting of the males and the negative response of many females. The basic courtship pattern of the male consists of orientation, followed by a display posture close to the female typically at the rear with his head under her wing tips, although males of a small number of species posture at the front of a female and still fewer at her side. The positions are species-specific. After assuming the display posture, the male engages in a number of complex, ritualized, and species-specific movements utilizing various parts of his body. Non-receptive females respond to the male's overtures by various avoidance patterns such as fleeing, kicking with their hind legs, depressing the tip of their abdomen toward the substrate, or ignoring his action. Where the female does not flee, males may court non-receptive females for long periods of time. A receptive female, however, accepts the male's courtship quite quickly, indicating acceptance by spreading her wings horizontally, extruding the ovipositor, and depressing her body toward the substrate.

Returning to the Hawaiian species, field and laboratory studies indicate that the males of many species patrol and defend a small but definite courting territory. These sites are not randomly distributed in the rain forests where the flies live, but each species has special preferences apparently determined by factors such as light, humidity, temperature, and spatial conditions. These territories are invariably close to where the individuals feed and females oviposit. Some of the spots, once located, are found to have male occupants for extended periods. Associated with this behaviour has been the evolutionary development of sexual dimorphism in males. In over fifty Hawaiian species studied from the behavioural point of view, male dimorphic structures have been found to be involved in courtship and appear on various parts of the body, but mainly on the appendages, for example, forelegs and sometimes middle legs, but there are also mouth-part modifications which may involve enlargement, wing modifications consisting of clouds or patches of pigmentation in various species, and antennal dimorphism.

Each male defends a limited area such as a single leaf, fern pinna, section of a fern frond stem, or portion of a tree trunk, and concurrently he advertises his presence visually or otherwise. Males of some species repeatedly drag the tip of their abdomens over the substrate and in so doing, deposit a thin film of liquid; others assume a ritualized posture and extrude and retract a bubble of liquid from their anal papillae. Defence of territory, usually against other males but also against non-receptive females, frequently involves physical contact between individuals, and the most vigorous fly, which is usually the largest, often appears to be the victor, the original possessor being displaced. It seems that sexually receptive females are attracted to the 'advertising' males and so seek them out. Both in the field and in the laboratory, males have repeatedly been observed first to assume an aggressive posture before an intruding female and then to court her. If she is non-receptive, the male becomes aggressive and drives her away. Females of these Hawaiian species lack the ability to engage in the repelling actions of extruding and elevating the tip of the abdomen which females of many other species utilize when males court them while they are feeding.

Therefore, in conclusion there are quantitative and qualitative differences distinguishing these Hawaiian species from those found elsewhere. Thus, they do not engage in courtship on the feeding site. All individuals at the feeding site are quiet, moving about slowly, and show no avoidance or antagonistic reactions to each other. During cloudy days of high humidity, a feeding site may be continuously occupied from early morning to late afternoon, but in periods of sunlight and low humidity there may be diurnal behaviour. On top of the basic drosophiloid behaviour, therefore, these Hawaiian species have superimposed a type of behaviour in which each male selects and defends a courting arena near the food and ovipositional site, and in the arena males advertise their presence and females are attracted to them. Concurrently, male sexual dimorphism with complex diverse and unique courtship patterns have evolved. To summarize, these particular Hawaiian species differ from non-Hawaiian species by territoriality, aggression, and advertising in males, associated with the spatial separation of feeding and courtship sites.

So far, genetic studies on these Hawaiian species are limited as compared with the cosmopolitan species which have been mainly discussed in this book. However, the extraordinary complexity of behavioural patterns of the Hawaiian species, with their associated morphological changes, should contribute greatly to our understanding of the evolutionary biology of the genus. Ecological factors are of prime importance in determining the location of territories, and behavioural factors in the defence of these territories. In the more cosmopolitan species there is evidence for the genetic basis of behaviour, and for various ecological parameters, although the behavioural parameters particularly, are far less complex than in the Hawaiian species. Thus, to understand the evolutionary biology of the genus an integration of behaviour, ecology, and genetics is essential, as are studies in various environments both in the laboratory and in the wild. Unfortunately many of those species which are rare and bizarre are difficult to culture in the laboratory, but an approach in which their habitat in the wild is studied, combined with attempts to breed and study them in the laboratory, should slowly bring more understanding. If we can comprehend the evolutionary biology of the genus *Drosophila* in depth, then our knowledge of evolution itself will be enhanced.

16

BEHAVIOUR, ECOLOGY, AND EVOLUTION

THE factors discussed in this book are all involved in the over-all fitness of an organism, and there is no *a priori* reason for regarding any factor, however subtle, as making no contribution to fitness. Indeed, if a factor makes no apparent contribution to fitness this may merely reflect our ignorance, since it seems impossible to prove this. Even if fitness factors are demonstrated, there is then the extreme difficulty of establishing associations with loci or blocks of loci in the laboratory, and the frequent near-impossibility of this in natural environments. A further problem, well illustrated by some of the examples quoted, is that the fitness of a particular genotype depends on the environment in which it is. In *D. pseudoobscura*, for example, population cages containing various karyotypes in pairs tend to give stable equilibria at 25°C, as the heterokaryotypes were found to be fitter than the corresponding homokaryotypes, and actual equilibrium frequencies corresponded quite well with the fitness estimates made. Thus, irrespective of the initial frequencies of the karyotypes, the equilibrium point was usually attained. However, at 16·5°C little change occurred in the frequencies in the population cages and at 22°C an intermediate situation arose, showing the dependence of the equilibrium on the environment, in this case temperature. Considering the general question of environment, an argument was presented for larger fitness differentials between genotypes at extreme environments than at optimal environments, which seems to be based on a relatively lower fitness of homozygotes than of the corresponding heterozygotes in extreme environments, as compared with the situation in more optimal environments. Explanations of this have been proposed at the molecular level for temperature sensitivity due to variations in thermal stability of enzymes at various temperatures. Although hypotheses along these lines were proposed a few years ago (Langridge 1962), the recent discovery of widespread polymorphism for enzyme variants adds weight to the speculation, but, even so, the study of genotypes controlling enzyme variants in extreme environments has been rather neglected. Especially in ectotherms, a dependence of fitness on environment would be expected, as in fact seems to occur in some of the cases investigated in *Drosophila*.

In addition, fitness estimates are only applicable to loci in a given population, since the genetic backgrounds between populations may vary, as is well shown by the breakdown of heterozygote advantage in between-population crosses for certain *D. pseudoobscura* karyotypes. In other words, within populations gene complexes are coadapted within and between chromosomes, but not between populations. Thus, to speak of fitness as a property applying to a specific gene or karyotype is rather spurious. This was probably first stressed in models developed by Mather (1943), showing how genes controlling a metrical trait may be arranged along the chromosome under various modes of selection, especially stabilizing selection. In particular, the models show that the theories and concepts of population genetics, of which fitness is one, should deal not with gene frequencies at individual loci, but with whole chromosomes, their recombination properties, and the effects of homozygosity of certain segments. It is gratifying that this concept is being stressed in recent literature much more than was the case a few years ago.

The dependence of fitness on the environment and on the whole genotype makes it therefore impossible to define fitness as an invariant parameter associated with a particular genotype or karyotype, which presents the problem of the biological meaning of the definition of fitness. Fitness can be defined as the average number of progeny left by the carriers of a given genotype, relative to the number of progeny left by the carriers of other genotypes. Thus, we must consider all the factors which contribute to this, which may comprise most, if not all, of the behavioural and ecological factors discussed in this book. It is unfortunate that in an experimental programme only a few fitness factors are normally measured, but one question that has been little explored concerns relationships between fitness factors. Relationships between population size, productivity, and longevity are discussed in Chapter 11, and in Chapter 9 associations between fertility and progeny numbers in crosses under high levels of larval competition are considered. At the behavioural level, there is evidence in *D. melanogaster* that males which mate first on the first occasion also copulate more often, more successfully, and leave more progeny (Chapter 5). For polymorphic inversions of *D. pseudoobscura*, it is known that the heterokaryotypes are often superior in innate capacity for increase, population size, productivity, egg to adult viability, and mating frequency compared with the corresponding homokaryotypes, especially at 25°C. However, the relative associations between all these components in a given population is an issue that has been relatively unexplored, but is of considerable importance in the study of over-all fitness of organisms.

A recent report (Prout 1971*a*, *b*), however, indicates one possible direction; he described an experimental system for estimating the components of fitness operating in the adult phase of the life cycle of *D. melanogaster*. The system entails the mating together of contrived genotypic mixtures, and the progeny allow for the estimation of the adult components of fitness for the sexes separately. He applied his system to fourth-chromosome recessive mutants, eyeless ey^2 and shaven sv^n, which were studied in repulsion phase so that

ey^2/sv^n flies are wild-type. The chromosome is short so that recombination is not a relevant complication. The components of fitness estimated were larval viability in each sex, and from adults two components were obtained, one representing female fecundity and the other representing male mating ability or virility. The major differential components were for the adult components. Thus ey^2ey^2 and ey^2/sv^n females were superior to sv^nsv^n. The male adult component showed strong superiority of ey^2/sv^n over both homozygotes, and the depressed fitness values of the latter two homozygotes varied with the female genotype to which the males were mated, indicating mating interactions of a frequency-dependent type. Compared with the adult fitness components, the larval fitness components were small. Prout stressed the need to define a small number of components of fitness which encompass the entire life cycle, and which are accessible for experimental evaluation. He tested (Prout 1971b) his fitness estimates by attempting to predict the performance of experimental populations segregating for these same mutants, by incorporating the estimates of fitness components into recurrence equations. This approach led to a reasonably good prediction. Although we are still very far from being able to reproduce what may happen in nature, if we are to understand it, studies of this type seem necessary in the laboratory, using a variety of genotypes and environments.

It is finally of interest that the male mating behaviour component of fitness turned out to be of such importance, as this agrees with the earlier experiments of Merrell (1953) who found gene frequency changes in populations to be predictable from mating behaviour variations (see §3.4). Similarly, Spiess and Langer (1964a) found substantial mating-speed differences between homokaryotypes of $D.$ $pseudoobscura$ derived from Mather, California, such that there was an association between rapid mating and the frequencies of the karyotypes at Mather (§4.1). Another experiment showing the importance of mating comes from Fulker's (1966) work (§5.3), where he found that strains where males copulated quickly, and most rapidly, led to the most offspring being produced. From this evidence, and certain other evidence in Chapters 4 and 5, it is tempting to postulate that mating behaviour could represent a component of fitness of great importance, and is therefore worthy of much more detailed study than it has received so far, especially in natural populations.

Given that certain genotypes do leave on the average more progeny than others—that is, they are fitter—this will lead to evolution by natural selection. The change in the genetic structure of a population as a function of its genetic variation was given an exact algebraic formulation by Fisher's (1930) fundamental theorem of natural selection, which states that the rate of increase in fitness of a population of organisms is equal to its genetic variance in fitness at that time. In general it can be assumed that the process of natural selection will lead to genotypes with greater adaptation to the environment. The adaptations so developed can take many forms, from the very general in cosmopolitan species to the highly specialized as in some of the species adapted to one host species only. Genetic architectures may vary from the highly polymorphic for chromosomal rearrangements to the monomorphic, going from the centre of a species distribution to the margin. Levels of ethological isolation may vary according to whether a sibling species is sympatric or allopatric. Adaptation to high temperatures and to desiccation is due to a complex of physiological acclimation, behavioural factors, and genetic change, the relative importance of the three modes of adaptation being species-specific. Many other cases of adaptation are described in the earlier chapters of this book, and all are the result of natural selection. Fisher's fundamental theorem is, however, no predictor of the path that evolution may take, since there is a considerable degree of opportunism about evolution.

Another factor, little discussed by Fisher, which may be of considerable importance in the beginning of a period of adaptive radiation is the actual genes present in the founding individuals if there are few of them. Experiments cited in Chapter 8 show the importance of the genes in founder females leading to characteristics of strains which persist for many generations. For scutellar chaeta number this association with the genes derived from the founder females has been shown to be stable over a number of temperatures (Hosgood and Parsons 1971). Therefore, when we look at the situation in the wild we see that the real problem is infinitely more complex than is implied by the fundamental theorem of natural selection, which in any case is based on the assumption of random mating (although some theoretical work has been carried out aimed at removing this restriction—see Crow and Kimura 1970).

In the extraordinary Hawaiian drosophilid fauna, well over 400 endemic species have been described, deriving ultimately from very few ancestors. Most species are endemic to single islands, yet the species groups are widespread, suggesting multiple proliferations. There is a striking karyotypic similarity among clusters of closely related species (Carson 1970), even though they may be morphologically and ethologically diverse. A likely situation that may have occurred in the Hawaiian drosophilids is that there was an initial founder individual which finding itself in an unoccupied niche underwent a population flush, defined by Carson (1968) as the ascending phase of population growth. Now during this phase of rapid increase, natural selection for survival is relaxed as compared with the normal situation, so that a breakdown of fitness is likely releasing a multitude of recombinant genotypes. At the crest, therefore, there would be a multitude of recombinants, many with relatively low fitness. At the same time, high mortality might be expected as competition for food and breeding space ensues. One inevitable result is dispersion and migration, a large amount of which might be random. Certain of these migrants would effect new colonizations, and in this the 'founder effect' of Mayr (1963) would be likely to be important, especially if the migrating individuals reached areas not previously occupied by the species. Some, but certainly not all, of these colonizations would survive, perhaps in quite low numbers and there would be active selection in the niches occupied, especially as they would normally be completely isolated from each other. In this context the data of Ehrman (1964) are of interest (§7.4). She found divergence, as measured by sexual isolation, merely as a result of geographical isolation as simulated by different population cages under essentially identical environments. If clear environmental divergence is superimposed, then divergence should be even more rapid. Carson considers that the extraordinary proliferation of species in places like Hawaii may indeed be related to population flushes and to chance effects following the subsequent crashes, associated with the role played by isolation resulting from the physical forces of the environment.

Looking at the situation in the wild as Fisher (1930) did is also instructive, even though he discussed chance events rather little. Thus, he wrote: 'If therefore an organism be really in any high degree adapted to the place it fills in its environment, this adaptation will be constantly menaced by any undirected agencies liable to cause changes to either party in the adaptation.' As Fisher pointed out, deterioration will occur genetically, by mutations and other undirected changes such as recombination. For most organisms, if the physical environment is changing it must be regarded with few exceptions as continuously deteriorating, in the sense that in a given environment adaptation will be continuously favoured by natural selection. A changing environment will be to the detriment of the current status because of the previous history of natural selection to maximize fitness in the existing environment. More important than the physical environment will probably be the changes in associated organisms, since as each organism increases in fitness for survival so too will its enemies, competitors, and predators. In this way there will be a continual evolutionary process made up of far-reaching interactions between organisms and environment, leading to the formation of new species and the extinction of others. Extinction will occur when a species finds itself in a position unable to adapt rapidly enough to changes around it, perhaps because it has evolved an inflexible genetic system. Thus, new species continually occur, old species are continually eliminated, but there may be bursts of speciation in the exploitation of a series of new and relatively unoccupied niches.

Environmental factors of a man-made type will probably assume progressively more importance with time. While most examples of evolutionary change due to man-made changes in the environment have come from species of more economic significance than *Drosophila*, the study of such phenomena in certain of the species of *Drosophila* is instructive, and shows some of the extremely rapid changes which may occur. The build-up of resistance to chemicals such as DDT is a good example, since DDT-resistant flies have a genome substantially different from DDT-sensitives. The rapid build-up of resistance is possible, presumably because before DDT appeared flies were not subjected to directional selection for resistance, as has occurred after its appearance. Therefore, responses to selection will be expected to be rapid, as in fact is the case. Laboratory experiments on directional selection for many traits not previously having a history of such selection frequently show rapid responses to directional selection for morphological, behavioural, and physiological traits alike. However, this is usually associated with a general decline in over-all fitness as assessed by increased sterility, reduced fecundity, lowered longevity, and poorer egg-hatchability and larva-to-adult viability. In other words, directional selection for a trait may lead to the line becoming unfit, and perhaps extinct. The effect of man on some species in the natural environment may well be similar, and chemical pollutants are likely to be as powerful selective forces as any, since they may have relatively direct and severe physiological effects on organisms, if interacting directly with various metabolic pathways.

In this discussion, we are really considering the adaptedness of populations of organisms (Chapter 11), by which is meant the ability of the population to survive and reproduce in its environments. This is an absolute measure, and we have seen the difficulty of its estimation in the laboratory. But to understand the processes of natural selection in leading to the various behavioural and ecological mechanisms discussed in the previous chapters, we must aim at formulating ways of assessing the over-all adaptedness of populations in nature. We are very distant from that goal, and it may be too complex to aim for at the present moment, but some components are slowly being understood. Fisher's fundamental theorem can presumably be extended to populations, in the form that 'the adaptedness of a population to the environments in which it lives increases at a rate proportional to the genetic variability present in the population' (Ayala 1969a). This, at the moment, is a qualitative rather than a quantitative statement, and perhaps with time we may hope for a quantification of it. However, one of the elements of greatest difficulty in this definition from the practical point of view is the issue of the environment.

In conclusion, the main text of the book is concerned with the various behavioural and ecological factors known to be of importance in leading to adaptation to the environment, regarding environment in the broadest possible sense as consisting of both living and non-living components. Although there are behavioural and ecological parts in the book the division is artificial, since many of the factors isolating populations and determining their distribution are a combination of ecological and behavioural components. In the shift into new adaptive zones with differing environments representing new ecological situations, habitat and food selection, which are mainly behavioural phenomena, are of major importance. Therefore, we see the importance of behaviour in initiating new evolutionary phenomena in a new niche, and its interaction with the ecology of the situation. The likely importance of behavioural phenomena in evolutionary change may help to explain the high level of variation of behavioural traits compared with morphological traits. Probably permanent morphological changes follow behavioural changes, and, as we have seen, there is an association between behavioural and morphological divergence. It is interesting, but perhaps not unexpected, that the most specialized of the Drosophilidae behaviourally seem to be the Hawaiian species, which are associated with substantial degrees of morphological and ecological specialization. Therefore, detailed studies of the genus going from the generalized to specialized species may slowly provide insight into the evolutionary biology of *Drosophila*, and the principles found may be applicable to other genera. If this proves to be so, a major aim of the book will have been achieved.

BIBLIOGRAPHY

AKIHAMA, T. (1968). Inheritance of the competitive ability and effects of its selection on agronomic characters. *Jap. J. Breed.* **18**, 12–14.

ANDERSON, W. W. (1968). Further evidence for coadaptation in crosses between geographic populations of *Drosophila pseudoobscura*. *Genet. Res.* **12**, 317–30.

—— (1969). Genetics of natural populations. XLI. The selection coefficients of heterozygotes for lethal chromosomes in *Drosophila* on different genetic backgrounds. *Genetics, Princeton* **62**, 827–36.

—— and EHRMAN, L. (1969). Mating choice in crosses between geographic populations of *Drosophila pseudoobscura*. *Am. Midl. Nat.* **81**, 47–53.

——, OSHIMA, C., WATANABE, T., DOBZHANSKY, T., *and* PAVLOVSKY, O. (1968). Genetics of natural populations. XXXIX. A test of the possible influence of two insecticides on the chromosomal polymorphism in *Drosophila pseudoobscura*. *Genetics, Princeton* **58**, 423–34.

ANDREWARTHA, H. G. *and* BIRCH, L. C. (1954). *The distribution and abundance of animals*. University of Chicago Press, Chicago and London.

AYALA, F. J. (1965*a*). Evolution of fitness in experimental populations of *Drosophila serrata*. *Science, N.Y.* **150**, 903–5.

—— (1965*b*). Relative fitness of populations of *Drosophila serrata* and *Drosophila birchii*. *Genetics, Princeton* **51**, 527–44.

—— (1966*a*). Evolution of fitness. I. Improvement in the productivity and size of irradiated populations of *Drosophila serrata* and *Drosophila birchii*. *Genetics, Princeton* **53**, 883–95.

—— (1966*b*). Dynamics of populations. I. Factors controlling population growth and population size in *Drosophila serrata*. *Am. Nat.* **100**, 333–44.

—— (1966*c*). Reversal of dominance in competing species of *Drosophila*. *Am. Nat.* **100**, 81–3.

—— (1968*a*). Genotype, environment, and population numbers. *Science, N.Y.* **162**, 1453–9.

—— (1968*b*). Environmental factors limiting the productivity and size of experimental populations of *Drosophila serrata* and *Drosophila birchii*. *Ecology* **49**, 562–5.

—— (1969*a*). An evolutionary dilemma: fitness of genotypes versus fitness of populations. *Can. J. Genet. Cytol.* **11**, 439–56.

—— (1969*b*). Genetic polymorphism and interspecific competitive ability in *Drosophila*. *Genet. Res.* **14**, 95–102.

—— (1969*c*). Evolution of fitness. IV. Genetic evolution of interspecific competitive ability in *Drosophila*. *Genetics, Princeton* **61**, 737–47.

—— (1969*d*). Experimental invalidation of the principle of competitive exclusion. *Nature, Lond.* **224**, 1076–9.

—— (1970*a*). Competition, coexistence, and evolution. In *Essays in evolution and genetics in honor of Theodosius Dobzhansky* (eds. M. K. Hecht *and* W. C. Steere), pp. 121–58. Appleton-Century-Crofts, New York.

—— (1970*b*). Population fitness of geographic strains of *Drosophila serrata* as measured by interspecific competition. *Evolution, Lancaster, Pa.* **24**, 483–94.

BAKKER, K. (1961). An analysis of factors which determine success in competition for food among larvae of *Drosophila melanogaster*. *Archs. néerl. Zool.* **14**, 2, 200–81.

BAND, H. T. (1963). Genetic structure of populations. II. Viabilities and variances of heterozygotes in constant and fluctuating environments. *Evolution, Loncaster, Pa.* **17**, 307–19.

—— and IVES, P. T. (1968). Genetic structure of populations. IV. Summer environmental variables and lethal and semilethal frequencies in a natural population of *Drosophila melanogaster*. *Evolution, Lancaster, Pa.* **22**, 633–41.

BARKER, J. S. F. (1962a). Studies of selective mating using the yellow mutant of *Drosophila melanogaster*. *Genetics, Princeton* **47**, 623–40.

—— (1962b). Sexual isolation between *Drosophila melanogaster* and *Drosophila simulans*. *Am. Nat.* **96**, 105–15.

—— (1967). Factors affecting sexual isolation between *Drosophila melanogaster* and *Drosophila simulans*. *Am. Nat.* **101**, 277–87.

—— and PODGER, R. N. (1970). Interspecific competition between *Drosophila melanogaster* and *Drosophila simulans*. Effects of larval density and short-term adult starvation on fecundity, egg hatchability and adult viability. *Ecology* **51**, 855–64.

BARNES, B. W. (1968). Stabilising selection in *Drosophila melanogaster*. *Heredity, Lond.* **23**, 433–42.

BASTOCK, M. (1956). A gene mutation which changes a behavior pattern. *Evolution, Lancaster, Pa.* **10**, 421–39.

BATEMAN, A. J. (1949). Analysis of data on sexual isolation. Evolution **3**, 174–7.

—— (1950). Is gene dispersion normal? *Heredity, Lond.* **4**, 353–63.

BATTAGLIA, B. *and* SMITH, H. (1961). The Darwinian fitness of polymorphic and monomorphic populations of *Drosophila pseudoobscura* at 16°C. *Heredity, Lond.* **16**, 475–84.

BEARDMORE, J. A. (1961). Diurnal temperature fluctuation and genetic variance in *Drosophila* populations. *Nature, Lond.* **189**, 162–3.

—— (1963). Mutual facilitation and the fitness of polymorphic populations. *Am. Nat.* **97**, 69–74.

—— (1970). Ecological factors and the variability of gene-pools in *Drosophila*. In *Essays in evolution and genetics in honor of Theodosius Dobzhansky* (eds. M. K. Hecht *and* W. C. Steere), pp. 299–314. Appleton-Century-Crofts, New York.

——, DOBZHANSKY, T., *and* PAVLOVSKY, O. A. (1960). An attempt to compare the fitness of polymorphic and monomorphic experimental populations of *Drosophila pseudoobscura*. *Heredity, Lond.* **14**, 19–33.

BECKER, H. J. (1970). The genetics of chemotaxis in *Drosophila melanogaster*: Selection for repellent insensitivity. *Molec. Gen. Genetics* **107**, 194–200.

BENNET-CLARK, H. C. *and* EWING, A. W. (1969). Pulse interval as a critical parameter in the courtship song of *Drosophila melanogaster*. *Anim. Behav.* **17**, 755–9.

BENZER, S. (1967). Behavioural mutants of *Drosophila* isolated by countercurrent distribution. *Proc. natnl. Acad. Sci. U.S.A.* **58**, 1112–19.

BERGER, E. M. (1970). A comparison of gene-enzyme variation between *Drosophila melanogaster* and *D. simulans*. *Genetics, Princeton* **66**, 677–83.

—— (1971). A temporal survey of allelic variation in natural and laboratory populations of *Drosophila melanogaster*. *Genetics, Princeton* **67**, 121–36.

BIBLIOGRAPHY

BIRCH, L. C. (1955). Selection in *Drosophila pseudoobscura* in relation to crowding. *Evolution, Lancaster, Pa.* **9**, 389–99.
—— (1957). The meanings of competition. *Am. Nat.* **91**, 5–18.
—— DOBZHANSKY, T., ELLIOTT, P. O., and LEWONTIN, R. C. (1963). Relative fitness of geographic races of *Drosophila serrata*. *Evolution, Lancaster, Pa.* **17**, 72–83.
BLIGHT, W. C. and ROMANO, A. (1953). Notes on a breeding site of *Drosophila americana* near St. Louis, Missouri. *Am. Nat.* **87**, 111–12.
BODMER, W. F. and PARSONS, P. A. (1960). The initial progress of new genes with various genetic systems. *Heredity, Lond.* **15**, 283–99.
—— —— (1962). Linkage and recombination in evolution. *Adv. Genet.* **11**, 1–100.
BOROWSKY, R. (1971). Principle of competitive exclusion and *Drosophila*. *Nature, Lond.* **230**, 409–10.
BÖSIGER, E. (1960). Sur le rôle de la sélection sexuelle dans l'évolution. *Experientia* **16**, 270–3.
BREESE, E. L. and MATHER, K. (1960). The organisation of polygenic activity within a chromosome in *Drosophila*. II. Viability. *Heredity, Lond.* **14**, 375–99.
BRNCIC, D. (1954). Heterosis and the integration of the genotype in geographic populations of *Drosophila pseudoobscura*. *Genetics, Princeton* **39**, 77–88.
—— (1961). Integration of the genotype in geographic populations of *Drosophila pavani*. *Evolution, Lancaster, Pa.* **15**, 92–7.
—— (1962). Chromosomal structure of populations of *Drosophila flavopilosa* studied in larvae collected in their natural breeding sites. *Chromosoma* **13**, 183–95.
—— (1966). Ecological and cytogenetic studies of *Drosophila flavopilosa*, a neotropical species living in *Cestrum* flowers. *Evolution, Lancaster, Pa.* **20**, 16–29.
—— (1968). The effects of temperature on chromosomal polymorphism of *Drosophila flavopilosa* larvae. *Genetics, Princeton* **59**, 427–32.
—— (1969). Long-term changes in chromosomally polymorphic laboratory stocks of *Drosophila pavani*. *Evolution, Lancaster, Pa.* **23**, 502–8.
—— (1970). Studies on the evolutionary biology of Chilean species of *Drosophila*. In *Essays in evolution and genetics in honor of Theodosius Dobzhansky* (eds. M. K. Hecht and W. C. Steere), pp. 401–36. Appleton-Century-Crofts, New York.
—— and KOREF-SANTIBAEÑZ, S. (1964). Mating activity of homo- and heterokaryotypes in *Drosophila pavani*. *Genetics, Princeton* **49**, 585–91.
BROADHURST, P. L. (1960). Experiments in psychogenetics. Applications of biometrical genetics to the inheritance of behaviour. Experiments in Personality, vol. 1. (ed. H. J. Eysenck), pp. 1–102. Routledge and Kegan Paul, London.
—— (1967). An introduction to the diallel cross. In *Behavior-genetic analysis* (ed. J. Hirsch), pp. 287–304. McGraw-Hill, New York.
BROWN, R. G. B. (1965). Courtship behaviour in the *Drosophila obscura* group. Part II. Comparative studies. *Behaviour* **25**, 281–323.
BUDNIK, M., BRNCIC, D. and KOREF-SANTIBAÑEZ, S. (1971). The effects of crowding on chromosomal polymorphism of *Drosophila pavani*. *Evolution, Lancaster, Pa.* **25**, 410–19.
BURLA, H., CUNHA, A. B. DA, CAVALCANTI, A. G. L., DOBZHANSKY, T., and PAVAN, C. (1950). Population density and dispersal rates in Brazilian *Drosophila willistoni*. *Ecology* **31**, 393–404.

BURNETT, B., CONNOLLY, K., and BECK, J. (1968). Phenogenetic studies on visual acuity in *Drosophila melanogaster*. *J. Insect. Physiol.* **14**, 855–60.

BURNS, J. M. and JOHNSON, F. M. (1971). Esterase polymorphism in the butterfly *Hemiargus isola*. Stability in a variable environment. *Proc. natnl. Acad. Sci. U.S.A.* **68**, 34–7.

CARMODY, G., COLLAZO, A. D., DOBZHANSKY, T., EHRMAN, L., JAFFREY, I. S., KIMBALL, S., OBREBSKI, S., SILAGI, S., TIDWELL, T., and ULLRICH, R. (1962). Mating preferences and sexual isolation within and between the incipient species of *Drosophila paulistorum*. *Am. Midl. Nat.* **68**, 67–82.

CARPENTER, F. W. (1905). The reactions of the pomace fly (*Drosophila ampelophila* Loew) to light, gravity, and mechanical stimulation. *Am. Nat.* **39**, 157–171.

CARSON, H. L. (1951). Breeding sites of *Drosophila pseudoobscura* and *Drosophila persimilis* in the transition zone of the Sierra Nevada. *Evolution, Lancaster, Pa.* **5**, 91–6.

—— (1958*a*). The population genetics of *Drosophila robusta*. *Adv. Genet.* **9**, 1–40.

—— (1958*b*). Response to selection under different conditions of recombination in *Drosophila*. *Cold Spring Harb. Symp. quant. Biol.* **23**, 291–306.

—— (1961*a*). Relative fitness of genetically open and closed experimental populations of *Drosophila robusta*. *Genetics, Princeton* **46**, 553–67.

—— (1961*b*). Heterosis and fitness in experimental populations of *Drosophila melanogaster*. *Evolution, Lancaster, Pa.* **15**, 496–509.

—— (1967). The association between *Drosophila carcinophila* Wheeler and its host, the land crab *Gecarcinus ruricola* (L). *Am. Midl. Nat.* **78**, 324–43.

—— (1968). The population flush and its genetic consequences. In *Population biology and evolution* (ed. R. C. Lewontin). Syracuse University Press, Syracuse.

—— (1970). Chromosome tracers of the origins of species. *Science, N.Y.* **168**, 1414–18.

——, HARDY, D. E., SPIETH, H. T., and STONE, W. S. (1970). The evolutionary biology of the Hawaiian Drosophilidae. In *Essays in evolution and genetics in honor of Theodosius Dobzhansky* (eds. M. K. Hecht and W. C. Steere), pp. 437–543. Appleton-Century-Crofts, New York.

——, KNAPP, E. P. and PHAFF, H. J. (1956). Studies on the ecology of *Drosophila* in the Yosemite region of California III. The yeast flora of the natural breeding sites of some species of *Drosophila*. *Ecology* **37**, 538–44.

—— and WASSERMAN, M. (1965). A widespread chromosomal polymorphism in a widespread species, *Drosophila buzzatii*. *Am. Nat.* **99**, 111–15.

—— and WHEELER, M. R. (1967). *Drosophila endobranchia*, a new Drosophilid associated with land crabs in the West Indies. *Ann. ent. Soc. Am.* **61**, 675–8.

CASPARI, E. (1968). Genetic endowment and environment in the determination of human behavior: biological viewpoint. *Am. Educational Res. J.* **5**, 43–55.

CAVALLI, L. L. (1952). An analysis of linkage in quantitative inheritance. In *Quantitative inheritance* (eds. E. C. Reeve and C. H. Waddington), pp. 135–44. H.M.S.O., London.

CAVALLI-SFORZA, L. (1962). The distribution of migration distances: models, and applications to genetics. In *Les déplacements humains—aspects méthodologiques de leur mesure* (ed. J. Sutter), pp. 139–58. Hachette, Paris.

CHETVERIKOV, S. S. (1926). On certain features of the evolutionary process from the viewpoint of modern genetics. *Zh. éksp. Biol. Med.* **2**, 3–54.

CHURCH, R. B. and ROBERTSON, F. W. (1966). Biochemical analysis of genetic differences in the growth of *Drosophila*. *Genet. Res.* **7**, 383–407.

CLAUSEN, J., KECK, D. D., and HIESEY, W. M. (1940). Experimental studies on the nature of species. I. Effect of varied environments on western North American plants. *Publs. Carnegie Instn.* No. 520, 1–452.

—— —— —— (1948). Experimental studies on the nature of species. III. Environmental responses of climatic races of *Achillea*. *Publs. Carnegie Instn.* No. 581, 1–129.

CLAYTON, F. E. (1968). Metaphase configurations in species of the Hawaiian Drosophilidae. *Univ. Tex. Publs.*, No. 6818, 263–78.

CONNOLLY, K. (1966). Locomotor activity in *Drosophila*. II. Selection for active and inactive strains. *Anim. Behav.* **14**, 444–9.

—— (1968). The social facilitation of preening behaviour in *Drosophila melanogaster*. *Anim. Behav.* **16**, 385–91.

——, BURNETT, B. and SEWELL, D. (1969). Selective mating and eye pigmentation: an analysis of the visual component in the courtship behavior of *Drosophila melanogaster*. *Evolution, Lancaster, Pa.* **23**, 548–59.

COOCH, F. G. and BEARDMORE, J. A. (1959). Assortative mating and reciprocal difference in the Blue Snow Goose complex. *Nature, Lond.* **183**, 1833–4.

COOPER, D. M. and DOBZHANSKY, T. (1956). Studies on the ecology of *Drosophila* in the Yosemite region of California. I. The occurrence of species of *Drosophila* in different life zones and at different seasons. *Ecology* 37, 526–33.

CORDEIRO, A. R., TOWNSEND, J. I., PETERSEN, J. A., and JAEGER, E. C. (1958). Genetics of Southern marginal populations of *Drosophila willistoni*. *Proc. X Int. Conf. Genet.*, vol. 2, 58–9.

CROSSLEY, S. and ZUILL, E. (1970). Courtship behaviour of some *Drosophila melanogaster* mutants. *Nature, Lond.* **225**, 1064–5.

CROW, J. F. (1954). Analysis of a DDT-resistant strain of *Drosophila*. *J. econ. Ent.* **47**, 393–8.

—— (1957). Genetics of insect resistance to chemicals. *A. Rev. Ent.* **2**, 227–46.

—— (1958). Some possibilities for measuring selection intensities in man. *Hum. Biol.* **30**, 1–13.

—— (1960). Genetics of insecticide resistance: General considerations. *Misc. Publs. Ent. Soc. Am.* **2** (1), 69–74.

—— (1968). Some analyses of hidden variability in *Drosophila* populations. In *Population biology and evolution* (ed. R. C. Lewontin), pp. 71–86. Syracuse University Press, Syracuse.

—— and KIMURA, M. (1970). *An introduction to population genetics theory*. Harper and Row, New York.

CRUMPACKER, D. W. and MARINKOVIC, D. (1967). Preliminary evidence of cold temperature resistance in *Drosophila pseudoobscura*. *Am. Nat.* **101**, 505–514.

CUNHA, A. B. DA (1951). Modification of the adaptive values of chromosomal types in *Drosophila pseudoobscura* by nutritional variables. *Evolution, Lancaster, Pa.* **5**, 395–404.

—— (1955). Chromosomal polymorphism in the Diptera. *Adv. Genet.* **7**, 93–138.

—— and DOBZHANSKY, T. (1954). A further study of chromosomal polymorphism in *Drosophila willistoni* in its relation to the environment. *Evolution, Lancaster, Pa.* **8**, 119–34.

CUNHA, A. B. DA, PAVLOVSKY, O., *and* SPASSKY, B. (1959). Genetics of natural populations. XXVIII. Supplementary data on the chromosomal polymorphism in *Drosophila willistoni* in its relation to the environment. *Evolution, Lancaster, Pa.* **13**, 389–404.

—— —— *and* SOKOLOFF, A. (1951). On food preferences of sympatric species of *Drosophila*. *Evolution, Lancaster, Pa.* **5**, 97–101.

DARLINGTON, C. D. (1937). *Recent advances in cytology* (2nd edn.). Churchill, London.

DAWOOD, M. M. *and* STRICKBERGER, M. W. (1969). The effect of larval interaction on viability in *Drosophila melanogaster*. III. Effects of biotic residues. *Genetics, Princeton* **63**, 213–20.

DEERY, B. J. *and* PARSONS, P. A. (1972a). Variations in the resistance of natural populations of *Drosophila* to phenyl-thio-carbamide (PTC). *Egypt. J. Genet. Cytol.* **1**, 13–17.

—— (1972b). Ether resistance in *Drosophila melanogaster*. Theoret. Appl. Genet.

DEMEREC, M. (ed.) (1955). *Biology of* Drosophila. Hafner, New York.

DOBZHANSKY, T. (1944). Experiments on sexual isolation in *Drosophila*. III. Geographic strains of *Drosophila sturtevanti*. *Proc. natn. Acad. Sci. U.S.A.* **30**, 335–9.

—— (1947a). Genetics of natural populations. XIV. A response of certain gene arrangements in the third chromosome of *Drosophila pseudoobscura* to natural selection. *Genetics, Princeton* **32**, 142–60.

—— (1947b). Adaptive changes induced by natural selection in wild populations of *Drosophila*. *Evolution, Lancaster, Pa.* **1**, 1–16.

—— (1947c). A directional change in the genetic constitution of a natural population of *Drosophila pseudoobscura*. *Heredity, Lond.* **1**, 53–64.

—— (1948a). Genetics of natural populations. XVIII. Experiments on chromosomes of *Drosophila pseudoobscura* from different geographic regions. *Genetics, Princeton* **33**, 588–602.

—— (1948b). Genetics of natural populations. XVI. Altitudinal and seasonal changes produced by natural selection in certain populations of *Drosophila pseudoobscura* and *Drosophila persimilis*. *Genetics, Princeton* **33**, 158–76.

—— (1950). Genetics of natural populations. XIX. Origin of heterosis through natural selection in populations of *Drosophila pseudoobscura*. *Genetics, Princeton* **35**, 288–302.

—— (1951). *Genetics and the Origin of Species* (3rd ed., revised) Columbia University Press, New York.

—— (1957a). Mendelian populations as genetic systems. *Cold Spring Harb. Symp. quant. Biol.* **22**, 385–93.

—— (1957b). Genetics of natural populations. XXVI. Chromosomal variability in island and continental populations of *Drosophila willistoni* from Central America and the West Indies. *Evolution, Lancaster, Pa.* **11**, 280–93.

—— (1959). Evolution of genes and genes in evolution. *Cold Spring Harb. Symp. quant. Biol.* **24**, 15–30.

—— (1968). On some fundamental concepts of Darwinian biology. *Evolutionary Biol.* **2**, 1–34.

——, ANDERSON, W. W., *and* PAVLOVSKY, O. (1966). Genetics of natural populations. XXXVIII. Continuity and change in populations of *Drosophila pseudoobscura* in Western United States. *Evolution, Lancaster, Pa.* **20**, 418–27.

—— —— ——, SPASSKY, B., *and* WILLS, C. J. (1964). Genetics of natural populations. XXXV. A progress report on genetic changes in populations of *Drosophila pseudoobscura* in the American Southwest. *Evolution, Lancaster, Pa.* **18**, 164–76.

——, BURLA, H., *and* CUNHA, A. B. DA (1950). A comparative study of chromosomal polymorphism in sibling species of the *willistoni* group of *Drosophila*. *Am. Nat.* **84**, 229–46.

——, COOPER, D. M., PHAFF, H. J., KNAPP, E. P., *and* CARSON, H. L. (1956). Studies of the ecology of *Drosophila* in the Yosemite region of California. IV. Differential attraction of species of *Drosophila* to different species of yeasts. *Ecology* **37**, 544–50.

—— *and* CUNHA, A. B. DA (1955). Differentiation of nutritional preferences in Brazilian species of *Drosophila*. *Ecology* **36**, 34–9.

——, EHRMAN, L., *and* KASTRITSIS, C. D. (1968). Ethological isolation between sympatric and allopatric species of the *obscura* group of *Drosophila*. *Anim. Behav.* **16**, 79–87.

——, HUNTER, A. S., PAVLOVSKY, O., SPASSKY, B., *and* WALLACE, B. (1963). Genetics of natural populations. XXXI. Genetics of an isolated marginal population of *Drosophila pseudoobscura*. *Genetics, Princeton* **48**, 91–103.

—— *and* LEVENE, H. (1951). Development of heterosis through natural selection in experimental populations of *Drosophila pseudoobscura*. *Am. Nat.* **85**, 247–64.

—— ——, SPASSKY, B., *and* SPASSKY, N. (1959). Release of genetic variability through recombination. III. *Drosophila prosaltans*. *Genetics, Princeton* **44**, 75–92.

——, LEWONTIN, R. C., *and* PAVLOVSKY, O. (1964). The capacity for increase in chromosomally polymorphic and monomorphic populations of *Drosophila pseudoobscura*. *Heredity, Lond.* **19**, 597–614.

—— *and* MAYR, E. (1944). Experiments on sexual isolation in *Drosophila*. I. Geographic strains of *Drosophila willistoni*. *Proc. natnl. Acad. Sci. U.S.A.* **30**, 238–44.

—— *and* PAVAN, C. (1950). Local and seasonal variations in relative frequencies of species of *Drosophila* in Brazil. *J. Anim. Ecol.* **19**, 1–14.

—— *and* PAVLOVSKY, O. (1953). The indeterminate outcome of certain experiments on *Drosophila* populations. *Evolution, Lancaster, Pa.* **7**, 198–210.

—— —— (1958). Interracial hybridization and breakdown of coadapted gene complexes in *Drosophila paulistorum* and *Drosophila willistoni*. *Proc. natn. Acad. Sci. U.S.A.* **44**, 622–9.

—— —— (1961). A further study of fitness of chromosomally polymorphic and monomorphic populations of *Drosophila pseudoobscura*. *Heredity, Lond.* **16**, 169–79.

—— —— (1967). Experiments on the incipient species of the *Drosophila paulistorum* complex. *Genetics, Princeton* **55**, 141–56.

—— —— *and* EHRMAN, L. (1969). Transitional populations of *Drosophila paulistorum*. *Evolution, Lancaster, Pa.* **23**, 482–92.

—— ——, SPASSKY, B., *and* SPASSKY, N. (1955). Genetics of natural populations. XXIII. Biological role of deleterious recessives in populations of *Drosophila pseudoobscura*. *Genetics, Princeton* **40**, 781–96.

—— *and* SPASSKY, B. (1959). *Drosophila paulistorum*, a cluster of species *in statu nascendi*. *Proc. natn. Acad. Sci. U.S.A.* **45**, 419–28.

DOBZHANSKY, T. *and* SPASSKY, B. (1962). Selection for geotaxis in monomorphic and polymorphic populations of *Drosophila pseudoobscura*. *Proc. natn. Acad. Sci. U.S.A.* **48**, 1704–12.

—— —— (1968). Genetics of natural populations. XL. Heterotic and deleterious effects of recessive lethals in populations of *Drosophila pseudoobscura*. *Genetics, Princeton* **59**, 411–25.

—— —— (1969). Artificial and natural selection for two behavioral traits in *Drosophila pseudoobscura*. *Proc. natn. Acad. Sci. U.S.A.* **62**, 75–80.

—— *and* STREISINGER, G. (1944). Experiments on sexual isolation in *Drosophila*. II. Geographic strains of *Drosophila prosaltans*. *Proc. natn. Acad. Sci. U.S.A.* **30**, 340–5.

—— *and* WRIGHT, S. (1941). Genetics of natural populations. V. Relations between mutation rate and accumulation of lethals in populations of *Drosophila pseudoobscura*. *Genetics, Princeton* **26**, 23–51.

—— —— (1943). Genetics of natural populations. X. Dispersion rates in *Drosophila pseudoobscura*. *Genetics, Princeton* **28**, 304–40.

—— —— (1947). Genetics of natural populations. XV. Rate of diffusion of a mutant gene through a population of *Drosophila pseudoobscura*. *Genetics, Princeton* **32**, 303–24.

DUBININ, N. P. *and* TINIAKOV, G. G. (1945). Seasonal cycles and the concentration of inversions in populations of *Drosophila funebris*. *Am. Nat.* **79**, 570–2.

—— —— (1946*a*). Structural chromosome variability in urban and rural populations of *Drosophila funebris*. *Am. Nat.* **80**, 393–6.

—— —— (1946*b*). Natural selection and chromosomal variability in populations of *Drosophila funebris*. *J. Hered.* **37**, 39–44.

—— —— (1946*c*). Inversion gradients and natural selection in ecological races of *Drosophila funebris*. *Genetics, Princeton* **31**, 537–45.

—— —— (1947). Inversion gradients and selection in ecological races of *Drosophila funebris*. *Am. Nat.* **81**, 148–53.

EHRMAN, L. (1960). The genetics of hybrid sterility in *Drosophila paulistorum*. *Evolution, Lancaster, Pa.* **14**, 212–23.

—— (1961). The genetics of sexual isloation in *Drosophila paulistorum*. *Genetics, Princeton* **46**, 1025–38.

—— (1964). Genetic divergence in M. Vetukhiv's experimental populations of *Drosophila pseudoobscura*. I. Rudiments of sexual isolation. *Genet. Res.* **5**, 150–7.

—— (1965). Direct observation of sexual isolation between allopatric and between sympatric strains of the different *Drosophila paulistorum* races. *Evolution, Lancaster, Pa.* **19**, 459–64.

—— (1966). Mating success and genotype frequency in *Drosophila*. *Anim. Behav.* **14**, 332–9.

—— (1967). Further studies on genotype frequency and mating success in *Drosophila*. *Am. Nat.* **101**, 415–24.

—— (1968. Frequency dependence of mating success in *Drosophila pseudoobscura*. *Genet. Res.* **11**, 135–40.

—— (1969). The sensory basis of mate selection in *Drosophila*. *Evolution, Lancaster, Pa.* **23**, 59–64.

—— (1970*a*). The mating advantage of rare males in *Drosophila*. *Proc. natn. Acad. Sci. U.S.A.* **65**, 345–8.

—— (1970*b*). A release experiment testing the mating advantage of rare *Drosophila* males. *Behav. Sci.* **15**, 363–5.

—— and PETIT, C. (1968). Genotype frequency and mating success in the *willistoni* species group of *Drosophila*. *Evolution, Lancaster, Pa.* **22**, 649–58.

——, SPASSKY, B., PAVLOVSKY, O., and DOBZHANSKY, T. (1965). Sexual selection, geotaxis, and chromosomal polymorphism in experimental populations of *Drosophila pseudoobscura*. *Evolution, Lancaster, Pa.* **19**, 337–46.

ELENS, A. A. (1957). Importance sélective des différences d'activité entre mâles *ebony* et *sauvage*, dans les populations artificielles de *Drosophila melanogaster*. *Experientia* **13**, 293–4.

—— (1958). Le rôle de l'hétérosis dans la compétition entre *ebony* et son allèle normal. *Experientia* **14**, 274–6.

—— and WATTIAUX, J. M. (1964). Direct observation of sexual isolation. *Drosoph. Inf. Serv.* **39**, 118–19.

ERLENMEYER-KIMLING, L., HIRSCH, J., and WEISS, J. M. (1962). Studies in experimental behavior genetics. III. Selection and hybridization analyses of individual differences in the sign of geotaxis. *J. comp. physiol. Psychol.* **55**, 722–31.

EWING, A. W. (1963). Attempts to select for spontaneous activity in *Drosophila melanogaster*. *Anim. Behav.* **11**, 369–78.

—— (1964). The influence of wing area on the courtship behaviour of *Drosophila melanogaster*. *Anim. Behav.* **12**, 316–20.

—— (1969). The genetic basis of sound production in *Drosophila pseudoobscura* and *D. persimilis*. *Anim. Behav.* **17**, 555–60.

FALCONER, D. S. (1960). *Introduction to quantitative genetics*. Oliver and Boyd, Edinburgh.

FINGERMAN, M. (1952). The role of the eye-pigments of *Drosophila melanogaster* in photic orientation. *J. exp. Zool.* **120**, 131–64.

FISHER, R. A. (1918). The correlation between relatives on the supposition of Mendelian inheritance. *Trans. R. Soc. Edinb.* **52**, 399–433.

—— (1922). On the dominance ratio. *Proc. R. Soc. Edinb.* **42**, 321–41.

—— (1930). *The genetical theory of natural selection*. Clarendon Press, Oxford.

FONTDEVILA, A. (1970). Genotype-temperature interaction in *Drosophila melanogaster*. I. Viability. *Genetica* **41**, 257–64.

FORD, E. B. (1964). *Ecological genetics*. Methuen, London.

FRANKLIN, I. and LEWONTIN, R. C. (1970). Is the gene the unit of selection? *Genetics, Princeton* **65**, 707–34.

FRASER, A. (1963). Variation of scutellar bristles in *Drosophila*. I. Genetic leakage. *Genetics, Princeton* **48**, 497–514.

——, ERWAY, L., and BRENTON, W. (1968). Variation of scutellar bristles in *Drosophila*. XIV. Effects of temperature and crowding. *Aust. J. biol. Sci.* **21**, 75–87.

FREIRE-MAIA, N. (1964). Chromosomal monomorphism in Brazilian and Argentine populations of *Drosophila simulans* and *Drosophila repleta*. *Genetics, Princeton* **50**, 1447–8.

FULKER, D. W. (1966). Mating speed in male *Drosophila melanogaster*: A psychogenetic analysis. *Science, N.Y.* **153**, 203–5.

FULLER, J. L. and THOMPSON, W. R. (1960). *Behavior genetics*. Wiley, New York.

FUTUYMA, D. J. (1970). Variation in genetic reponse to interspecific competition in laboratory populations of *Drosophila*. *Am. Nat.* **104**, 239–52.

GALE, J. S. (1964). Competition between three lines of *Drosophila melanogaster*. *Heredity, Lond.* **19**, 681–99.
GAUSE, G. F. (1934). *The struggle for existence*. Williams and Wilkins, Baltimore.
GEER, B. W. *and* GREEN, M. M. (1962). Genotype, phenotype and mating behavior of *Drosophila melanogaster*. *Am. Nat.* **96**, 175–81.
GEORGHIOU, G. P. (1965). Genetic studies on insecticide resistance. *Adv. Pest Control Res.* **6**, 171–230.
GIBSON, J. (1970). Enzyme flexibility in *Drosophila melanogaster*. *Nature, Lond.* **227**, 959–60.
—— *and* MIKLOVICH, R. (1971). Modes of variation in alcohol dehydrogenase in *Drosophila melanogaster*. *Experientia* **27**, 99–100.
—— *and* THODAY, J. M. (1962). Effects of disruptive selection. VI. A second chromosome polymorphism. *Heredity, Lond.* **17**, 1–26.
GLASS, B. *and* LI, C. C. (1953). The dynamics of racial intermixture—an analysis based on the American Negro. *Am. J. hum. Genet.* **5**, 1–20.
GRANT, B. *and* METTLER, L. E. (1969). Disruptive and stabilizing selection on the 'escape' behavior of *Drosophila melanogaster*. *Genetics, Princeton* **62**, 625–37.
GRIFFING, B. (1956). Concept of general and specific combining ability in relation to diallel crossing systems. *Aust. J. biol. Sci.* **9**, 463–93.
GRIGOLO, A. *and* OPPENOORTH, F. J. (1966). The importance of DDT-dehydrochlorinase for the effect of the resistance gene *kdr* in the housefly *Musca domestica* L. *Genetica* **37**, 159–70.
GROSSFIELD, J. (1966). The influence of light on the mating behavior of *Drosophila*. *Univ. Tex. Publs*. No. 6615, 147–76.
—— (1968). The relative importance of wing utilization in light dependent courtship in *Drosophila*. *Univ. Tex. Publs*. No. 6818, 147–56.
HADLER, N. M. (1964). Heritability and phototaxis in *Drosophila melanogaster*. *Genetics, Princeton* **50**, 1269–77.
HALDANE, J. B. S. (1937). The effect of variation on fitness. *Am. Nat.* **71**, 337–49.
—— (1957). The conditions for coadaptation in polymorphism for inversions. *J. Genet.* **55**, 218–25.
HARLAN, H. V. *and* MARTINI, M. L. (1938). The effect of natural selection in a mixture of barley varieties. *J. agric. Res.* **57**, 189–99.
HAY, D. A. (1972*a*). Genetical and maternal determinants of the activity and preening behaviour of *Drosophila melanogaster* reared in different environments. *Heredity, Lond.* **28**, 311–36.
—— (1972*b*). Genotype-environmental interaction in the activity and preening of *Drosophila melanogaster* (in preparation).
—— (1972*c*). Recognition by *Drosophila melanogaster* of individuals from other strains or cultures. Support for the role of olfactory cues in selective mating. Evolution (in press).
HAYMAN, B. I. (1954). The theory and analysis of diallel crosses. *Genetics, Princeton* **39**, 789–809.
HEED, W. B. (1968). Ecology of the Hawaiian Drosophilidae. *Univ. Tex. Publs*. No. 6818, 387–419.
—— *and* KIRCHER, H. W. (1965). Unique sterol in the ecology and nutrition of *Drosophila pachea*. *Science, N.Y.* **149**, 758–61.
HERSKOWITZ, I. H. (1951). A list of chemical substances studied for effects on *Drosophila*, with a bibliography. *Am. Nat.* **85**, 181–99.

HEUTS, M. J. (1948). Adaptive properties of carriers of certain gene arrangements in *Drosophila pseudoobscura. Heredity, Lond.* **2**, 63–75.

HIRSCH, J. (1962). Individual differences in behavior and their genetic basis. In *Roots of behavior* (ed. E. L. Bliss), pp. 3–23. Harper, New York.

—— (1963). Behavior genetics and individuality understood. *Science, N.Y.* **142**, 1436–42.

—— (1967). Behavior-genetic analysis at the chromosome level of organization. In *Behavior-genetic analysis* (ed. J. Hirsch), pp. 258–69. McGraw-Hill, New York.

—— and BOUDREAU, J. C. (1958). Studies in experimental behavior genetics. I. The heritability of phototaxis in a population of *Drosophila melanogaster. J. comp. physiol. Psychol.* **51**, 647–51.

—— and ERLENMEYER-KIMLING, L. (1962). Studies in experimental behavior genetics. IV. Chromosome analyses for geotaxis. *J. comp. physiol. Psychol.* **55**, 732–9.

HOENIGSBERG, H. F. (1968). An ecological situation which produced a change in the proportion of *Drosophila melanogaster* to *Drosophila simulans. Am. Nat.* **102**, 389–90.

——, CASTRO, L. E., GRANOBLES, L. A., and IDROBO, J. M. (1969). Population genetics in the American tropics. II. The comparative genetics of *Drosophila* in European and neo-tropical environments. *Genetica* **40**, 43–60.

——, CHEJNE, A. J., and HORTOBAGJI-GERMAN, E. (1966). Preliminary report on artificial selection towards sexual isolation in *Drosophila. Z. Tierpsychol.* **23**, 129–35.

——, GRANOBLES, L. A., and CASTRO, L. E. (1969). Population genetics in the American tropics. IV. Temporal changes effected in natural populations of *Drosophila melanogaster* from Colombia. *Genetica* **40**, 210–15.

HOSGOOD, S. M. W. and PARSONS, P. A. (1966). Differences between *D. simulans* and *D. melanogaster* in tolerances to laboratory temperatures. *Drosoph. Inf. Serv.* **41**, 176.

—— —— (1967a). Genetic heterogeneity among the founders of laboratory populations of *Drosophila melanogaster*. II. Mating behaviour. *Aust. J. Biol. Sci.* **20**, 1193–203.

—— —— (1967b). The exploitation of genetic heterogeneity among the founders of laboratory populations of *Drosophila* prior to directional selection. *Experientia* **23**, 1066–7.

—— —— (1968). Polymorphism in natural populations of *Drosophila* for the ability to withstand temperature shocks. *Experientia* **24**, 727–8.

—— —— (1971). Genetic heterogeneity among the founders of laboratory populations of *Drosophila*. IV. Scutellar chaetae in different environments. *Genetica* **42**, 42–52.

HUANG, S. L., SINGH, M., and KOJIMA, K-I. (1971). A study of frequency-dependent selection observed in the esterase-6 locus of *Drosophila melanogaster* using a conditioned media method. *Genetics, Princeton* **68**, 97–104.

HUTCHINSON, G. E. (1957). Concluding remarks. *Cold Sprng Harb. Symp. quant. Biol.* **22**, 415–27.

IVES, P. T. (1970). Further genetic studies of the South Amherst population of *Drosophila melanogaster. Evolution, Lancaster, Pa.* **24**, 507–18.

JACOBS, M. E. (1961). The influence of light on gene frequency changes in laboratory populations of ebony and non-ebony *Drosophila melanogaster. Genetics Princeton* **46**, 1089–95.

JINKS, J. L. (1954). The analysis of continuous variation in a diallel cross of *Nicotiana rustica* varieties. *Genetics, Princeton* **39**, 767–88.

JOHNSON, F. M. (1971). Isozyme polymorphisms in *Drosophila ananassae*: Genetic diversity among island populations in the South Pacific. *Genetics, Princeton* **68**, 77–95.

——, SCHAFFER, H. E., GILLASPY, J. F., and ROCKWOOD, E. S. (1969). Isozyme genotype-environment relationships in natural populations of the harvester ant, *Pogonomyrmex barbatus*, from Texas. *Biochem. Genet.* **3**, 429–50.

KALMUS, H. (1941). The resistance to desiccation of *Drosophila* mutants affecting body colour. *Proc. R. Soc.* B **130**, 185–201.

—— (1943). The optomotor responses of some eye mutants of *Drosophila. J. Genet.* **45**, 206–13.

—— (1945). Adaptive and selective responses of a population of *Drosophila melanogaster* containing e and $e+$ to differences in temperature, humidity and to selection for developmental speed. *J. Genet.* **47**, 58–63.

KAMBYSELLIS, M. P. *and* HEED, W. B. (1971). Studies of oogenesis in natural populations of Drosophilidae. I. Relation of ovarian development and ecological habitats of the Hawaiian species. *Am. Nat.* **105**, 31–49.

KAPLAN, W. D. *and* TROUT, W. E. (1969). The behavior of four neurological mutants of *Drosophila. Genetics, Princeton* **61**, 399–409.

KASTRITSIS, C. D. *and* DOBZHANSKY, T. (1967). *Drosophila pavlovskiana*, a race or a species? *Am. Midl. Nat.* **78**, 244–8.

KAUL, D. *and* PARSONS, P. A. (1965). The genotypic control of mating speed and duration of copulation in *Drosophila pseudoobscura. Heredity, Lond.* **20**, 381–92.

—— —— (1966). Competition between males in the determination of mating speed in *Drosophila pseudoobscura. Aust. J. biol. Sci.* **19**, 945–7.

KEARSEY, M. J. (1965a). The interaction of competition and food supply in two lines of *Drosophila melanogaster. Heredity, Lond.* **20**, 169–81.

—— (1965b). Cooperation among larvae of a wild-type strain of *Drosophila melanogaster. Heredity, Lond.* **20**, 309–12.

—— *and* BARNES, B. W. (1970). Variation for metrical characters in *Drosophila* populations. II. Natural selection. *Heredity, Lond.* **25**, 11–21.

—— *and* KOJIMA, K.-I. (1967). The genetic architecture of body weight and egg hatchability in *Drosophila melanogaster. Genetics, Princeton* **56**, 23–37.

KEMPTHORNE, O. (1957). *An introduction to genetic statistics.* Wiley, New York.

KESSLER, S. (1962). Courtship rituals and reproductive isolation between the races or incipient species of *Drosophila paulistorum. Am. Nat.* **96**, 117–21.

—— (1966). Selection for and against ethological isolation between *Drosophila pseudoobscura* and *Drosophila persimilis. Evolution, Lancaster, Pa.* **20**, 634–45.

—— (1968). The genetics of *Drosophila* mating behaviour. I. Organization of mating speed in *Drosophila pseudoobscura. Anim. Behav.* **16**, 485–91.

—— (1969). The genetics of *Drosophila* mating behavior. II. The genetic architecture of mating speed in *Drosophila pseudoobscura. Genetics, Princeton* **62**, 421–33.

KING, J. C. *and* SØMME, L. (1958). Chromosomal analysis of the genetic factors for resistance to DDT in two resistant lines of *Drosophila melanogaster. Genetics, Princeton* **43**, 577–93.

KIRCHER, H. W., HEED, W. B., RUSSELL, J. S., and GROVE, J. (1967). Serita cactus alkaloids: their significance to Sonoran desert *Drosophila* ecology. *J. Insect. Physiol.* **13**, 1869–74.

KNIGHT, G. R., ROBERTSON, A., and WADDINGTON, C. H. (1956). Selection for sexual isolation within a species. *Evolution, Lancaster, Pa.* **10**, 14–22.

KOEHN, R. K. (1969). Esterase heterogeneity: dynamics of a polymorphism. *Science, N.Y.* **163**, 943–4.

—— and RASMUSSEN, D. I. (1967). Polymorphic and monomorphic serum esterase heterogeneity in catostomid fish populations. *Biochem. Genet.* **1**, 131–44.

KOJIMA, K.-I., GILLESPIE, J., and TOBARI, Y. N. (1970). A profile of *Drosophila* species enzymes assayed by electrophoresis. I. Number of alleles, heterozygosities, and linkage disequilibrium in glucose-metabolizing systems and some other enzymes. *Biochem. Genet.* **4**, 627–37.

—— and LEWONTIN, R. C. (1970). Evolutionary significance of linkage and epistasis. In *Biomathematics* vol. 1 (ed. K.-I. Kojima), pp. 367–88. Springer-Verlag, Berlin.

—— and YARBROUGH, K. M. (1967). Frequency-dependent selection at the esterase 6-locus in *Drosophila melanogaster*. *Proc. natln. Acad. Sci. U.S.A.* **57**, 645–9.

KOOPMAN, K. F. (1950). Natural selection for reproductive isolation between *Drosophila pseudoobscura* and *Drosophila persimilis*. *Evolution, Lancaster, Pa.* **4**, 135–48.

KOREF-SANTIBAÑEZ, S. (1964). Reproductive isolation between the sibling species *Drosophila pavani* and *Drosophila gaucha*. *Evolution, Lancaster, Pa.* **18**, 245–51.

—— and WADDINGTON, C. H. (1958). The origin of sexual isolation between different lines within a species. *Evolution, Lancaster, Pa.* **12**, 485–93.

KOSUDA, K. (1971). Synergistic interaction between second and third chromosomes on viability of *Drosophila melanogaster*. *Jap. J. Genet.* **46**, 41–52.

KRIMBAS, C. B. (1959). Comparison of the concealed variability in *Drosophila willistoni* with that in *Drosophila prosaltans*. *Genetics, Princeton* **44**, 1359–69.

—— (1961). Release of genetic variability through recombination. VI. *Drosophila willistoni*. *Genetics, Princeton* **46**, 1323–34.

—— (1967). The genetics of *Drosophila subobscura* populations. III. Inversion polymorphism and climatic factors. *Molec. Gen. Genetics* **99**, 133–50.

KROMAN, R. A. and PARSONS, P. A. (1960). Genetic basis of two melanin inhibitors in *Drosophila melanogaster*. *Nature, Lond.* **186**, 411–12.

LACK, D. (1966). *Population studies of birds*. Clarendon Press, Oxford.

LANGRIDGE, J. (1962). A genetic and molecular basis for heterosis in *Arabidopsis* and *Drosophila*. *Am. Nat.* **96**, 5–27.

—— (1968). Thermal responses of mutant enzymes and temperature limits to growth. *Molec. Gen. Genetics* **103**, 116–26.

LEE, B. T. O. and PARSONS, P. A. (1968). Selection, prediction and response. *Biol. Rev.* **43**, 139–74.

LERNER, A. B. and FITZPATRICK, T. B. (1950). Biochemistry of melanin formation. *Physiol. Rev.* **30**, 91–126.

LEVENE, H. (1949). A new measure of sexual isolation. *Evolution, Lancaster, Pa.* **3**, 315–21.

—— (1959). Release of genetic variability through recombination. IV. Statistical theory. *Genetics, Princeton* **44**, 93–104.

LEVENE, H., PAVLOVSKY, O., and DOBZHANSKY, T. (1954). Interaction of the adaptive values in polymorphic experimental populations of *Drosophila pseudoobscura*. *Evolution, Lancaster, Pa.* **8**, 335–49.

LEVINE, R. P. (1952). Adaptive responses of some third chromosome types of *Drosophila pseudoobscura*. *Evolution, Lancaster, Pa.* **6**, 216–33.

LEVINS, R. (1968). *Evolution in changing environments: some theoretical explorations*. Princeton University Press, Princeton.

—— (1969). Thermal acclimation and heat resistance in *Drosophila* species. *Am. Nat.* **103**, 483–99.

LEWONTIN, R. C. (1955). The effects of population density and composition on viability in *Drosophila melanogaster*. *Evolution, Lancaster, Pa.* **9**, 27–41.

—— (1959). On the anomalous response of *Drosophila pseudoobscura* to light. *Am. Nat.* **93**, 321–8.

—— and HUBBY, J. L. (1966), A molecular approach to the study of genic heterozygosity in natural populations. II. Amount of variation and degree of heterozygosity in natural populations of *Drosophila pseudoobscura*. *Genetics, Princeton* **54**, 595–609.

—— and MATSUO, Y. (1963). Interaction of genotypes determining viability in *Drosophila bucksii*. *Proc. natnl. Acad. Sci. U.S.A.* **49**, 270–8.

LONG, T. (1970). Genetic effects of fluctuating temperature in populations of *Drosophila melanogaster*. *Genetics, Princeton* **66**, 401–16.

LOWE, C. H., HEED, W. B., and HALPERN, E. A. (1967). Supercooling of the saguaro species *Drosophila nigrospiracula* in the Sonoran Desert. *Ecology* **48**, 984–5.

MACARTHUR, R. H. (1965). Patterns of species diversity. *Biol. Rev.* **40**, 510–533.

—— (1968). The theory of the niche. In *Population biology and evolution* (ed. R. C. Lewontin), pp. 159–76. Syracuse University Press, Syracuse.

MACBEAN, I. T. and PARSONS, P. A. (1966). The genotypic control of the duration of copulation in *Drosophila melanogaster*. *Experientia* **22**, 101–2.

—— —— (1967). Directional selection for duration of copulation in *Drosophila melanogaster*. *Genetics, Princeton* **56**, 233–9.

MCFARQUHAR, A. M. and ROBERTSON, F. W. (1963). The lack of evidence for co-adaptation in crosses between geographical races of *Drosophila subobscura*. *Coll. Genet. Res.* **4**, 104–31.

MCKENZIE, J. A. and PARSONS, P. A. (1971). Variations in mating propensities in strains of *Drosophila melanogaster* with different scutellar chaeta numbers. *Heredity, Lond.* **26**, 313–22.

—— —— (1972). Alcohol tolerance: An ecological parameter in the relative success of *Drosophila melanogaster* and *Drosophila simulans*. Oecologia (in press).

MCWILLIAM, J. R. and GRIFFING, B. (1965). Temperature-dependent heterosis in maize. *Aust. J. biol. Sci.* **18**, 569–83.

MAINARDI, M. (1968). Gregarious oviposition and pheromones in *Drosophila melanogaster*. *Boll. Zool.* **35**, 135–6.

MALOGOLOWKIN-COHEN, C. H., SIMMONS, A. S., and LEVENE, H. (1965). A study of sexual isolation between certain strains of *Drosophila paulistorum*. *Evolution, Lancaster, Pa.* **19**, 95–103.

MANNING, A. (1959). The sexual behaviour of two sibling *Drosophila* species. *Behaviour* **15**, 123–45.

—— (1961). The effects of artificial selection for mating speed in *Drosophila melanogaster*. *Anim. Behav.* **9**, 82–92.

—— (1963). Selection for mating speed in *Drosophila melanogaster* based on the behaviour of one sex. *Anim. Behav.* **11**, 116–20.

—— (1966). Sexual behaviour. In *Insect behaviour* (ed. P. T. Haskell). *Symp. R. ent. Soc.* **3**, 59–68.

—— (1967). The control of sexual receptivity in female *Drosophila*. *Anim. Behav.* **15**, 239–50.

—— (1968a). *Drosophila* and the evolution of behaviour. In *Viewpoints in biology* (eds. J. D. Carthy and C. L. Duddington), pp. 125–69. Butterworths, London.

—— (1968b). The effects of artificial selection for slow mating in *Drosophila simulans*. I. The behavioural changes. *Anim. Behav.* **16**, 108–13.

MARINKOVIC, D., CRUMPACKER, D. W., *and* SALCEDA, V. M. (1969). Genetic loads and cold temperature resistance in *Drosophila pseudoobscura*. *Am. Nat.* **103**, 235–46.

MATHER, K. (1942). The balance of polygenic combinations. *J. Genet.* **43**, 309–336.

—— (1943). Polygenic inheritance and natural selection. *Biol. Rev.* **18**, 32–64.

—— (1949). *Biometrical genetics*. Methuen, London.

—— (1955). Polymorphism as an outcome of disruptive selection. *Evolution, Lancaster, Pa.* **9**, 52–61.

—— (1961). Competition and cooperation. *Symp. Soc. exp. Biol.* **15**, 264–81.

—— (1966). Variability and selection. *Proc. R. Soc.* B **164**, 328–40.

—— (1969). Selection through competition. *Heredity, Lond.* **24**, 529–40.

—— *and* COOKE, P. (1962). Differences in competitive ability between genotypes of *Drosophila*. *Heredity, Lond.* **17**, 381–407.

—— *and* HARRISON, B. J. (1949). The manifold effect of selection. *Heredity, Lond.* **3**, 1–52, 131–62.

—— *and* JINKS, J. L. (1971). *Biometrical genetics*. Chapman and Hall, London.

MAYHEW, S. H., KATO, S. K., BALL, F. M., *and* EPLING, C. (1966). Comparative studies of arrangements within and between populations of *Drosophila pseudoobscura*. *Evolution, Lancaster, Pa.* **20**, 646–62.

MAYNARD SMITH, J. (1956). Fertility, mating behaviour and sexual selection in *Drosophila subobscura*. *J. Genet.* **54**, 261–79.

—— (1962). Disruptive selection, polymorphism and sympatric speciation. *Nature, Lond.* **195**, 60–2.

MAYR, E. (1950). The role of the antennae in the mating behavior of female *Drosophila*. *Evolution, Lancaster, Pa.* **4**, 149–54.

—— (1963). *Animal species and evolution*. Harvard University Press, Cambridge.

—— *and* DOBZHANSKY, T. (1945). Experiments on sexual isolation in *Drosophila*. IV. Modification of the degree of isolation between *Drosophila pseudoobscura* and *Drosophila persimilis* and of sexual preferences in *Drosophila prosaltans*. *Proc. natnl. Acad. Sci. U.S.A.* **31**, 75–82.

MÉDIONI, J. (1959). Sur le rôle des yeux composés de *Drosophila melanogaster* Meigen dans la perception du proche ultraviolet: expériences sur le phototrophisme de races mutantes. *C. Séanc. Soc. Biol.* **153**, 164–7.

—— (1962). Contribution à l'étude psycho-physiologique et génétique du phototrophisme d'un insecte *Drosophila melanogaster* Meigen. *Bull. Psychol., Paris* **16**, (2), 8.

MEDIONI, J. (1963). La variabilité des comportements taxiques; ses principales conditions écologiques et organiques. *Ergebn. Biol.* **26**, 66–82.

MERRELL, D. J. (1949a). Selective mating in *Drosophila melanogaster*. *Genetics, Princeton* **34**, 370–89.

—— (1949b). Mating between two strains of *Drosophila melanogaster*. *Evolution, Lancaster, Pa.* **3**, 266–8.

—— (1951). Interspecific competition between *Drosophila funebris* and *Drosophila melanogaster*. *Am. Nat.* **85**, 159–69.

—— (1953). Selective mating as a cause of gene frequency changes in laboratory populations of *Drosophila melanogaster*. *Evolution, Lancaster, Pa.* **7**, 287–96.

—— (1965). Lethal frequency and allelism in DDT-resistant populations and their controls. *Am. Nat.* **99**, 411–17.

—— and UNDERHILL, J. C. (1956). Selection for DDT-resistance in inbred, laboratory, and wild stocks of *Drosophila melanogaster*. *J. econ. Ent.* **49**, 300–6.

MILLER, D. D. *and* WESTPHAL, N. J. (1967). Further evidence on sexual isolation within *Drosophila athabasca*. *Evolution, Lancaster, Pa.* **21**, 479–92.

MISRA, R. K. *and* REEVE, E. C. R. (1964). Clines in body dimensions in populations of *Drosophila subobscura*. *Genet. Res.* **5**, 240–56.

MIYOSHI, Y. (1961). On the resistibility of *Drosophila* to sodium chloride. I. Strain difference and heritability in *D. melanogaster*. *Genetics, Princeton* **46**, 935–45.

MOHN, N. *and* SPIESS, E. B. (1963). Cold resistance of karyotypes in *Drosophila persimilis* from Timberline of California. *Evolution, Lancaster, Pa.* **17**, 548–63.

MOORE, J. A. (1952a). Competition between *Drosophila melanogaster* and *Drosophila simulans*. I. Population cage experiments. *Evolution, Lancaster, Pa.* **6**, 407–20.

—— (1952b). Competition between *Drosophila melanogaster* and *Drosophila simulans*. II. The improvement of competitive ability through selection. *Proc. natnl. Acad. Sci. U.S.A.* **38**, 813–17.

MULLER, H. J. (1950). Our load of mutations. *Am. J. hum. Genet.* **2**, 111–76.

NAIR, P. S. (1969). Genetic load in *Drosophila robusta*. *Genetics, Princeton* **63**, 221–8.

NARISE, T. (1968). Migration and competition in *Drosophila*. I. Competition between wild and vestigial strains of *Drosophila melanogaster* in a cage and migration-tube population. *Evolution, Lancaster, Pa.* **22**, 301–6.

—— (1969). Migration and competition in *Drosophila*. II. Effect of genetic background on migratory behavior of *Drosophila melanogaster*. *Jap. J. Genet.* **44**, 297–302.

O'BRIEN, S. J., MACINTYRE, R. J., *and* FINE, W. (1968). A linkage disequilibrium between two gene-enzyme systems in an experimental population of *Drosophila melanogaster*. *Genetics, Princeton* **60**, 208–9.

—— —— (1969). An analysis of gene-enzyme variability in natural populations of *Drosophila melanogaster* and *D. simulans*. *Am. Nat.* **103**, 97–113.

O'DONALD, P. (1959). Possibility of assortive mating in the Arctic Skua. *Nature, Lond.* **183**, 1210–11.

OGAKI, M. *and* NAKASHIMA-TANAKA, E. (1966). Inheritance of radioresistance in *Drosophila*. I. *Mutation Res.* **3**, 438–43.

—— —— *and* MURAKAMI, S. (1967). Inheritance of ether resistance in *Drosophila melanogaster*. *Jap. J. Genet.* **42**, 387–94.

OGITA, Z. (1958). The genetical relation between resistance to insecticides in general and that to phenylthiourea (PTU) and phenylurea (PU) in *Drosophila melanogaster*. *Botyu-Kagaku* **23**, 188–205.

—— (1961). Genetical and biochemical studies on negatively correlated cross-resistance in *Drosophila melanogaster*. I. An attempt to reduce and increase insecticide-resistance in *D. melanogaster* by selection pressure. *Botyu-Kagaku* **26**, 7–18.

OHBA, S. (1967). Chromosomal polymorphism and capacity for increase under near-optimal conditions. *Heredity, Lond.* **22**, 169–86.

OKA, H. I. (1960). Variation in competitive ability among rice varieties. Phylogenetic differentiation in cultivated rice. XIX. *Jap. J. Breed.* **10**, 61–8.

OPPENOORTH, F. J. (1965). Biochemical genetics of insecticide resistance. *A. Rev. Ent.* **10**, 185–206.

OSHIMA, C. (1954). Genetical studies on DDT-resistance in populations of *Drosophila melanogaster*. *Botyu-Kagaku* **19**, 93–100.

—— (1969). Persistence of some recessive lethal genes in natural populations of *Drosophila melanogaster*. *Jap. J. Genet.* **44**, suppl. 1, 209–16.

—— and WATANABE, T. (1966). The effect of insecticide selection on experimental populations of *Drosophila pseudoobscura*. *Drosoph. Inf. Serv.* **41**, 140–1.

PARSONS, P. A. (1959*a*). Dependence of genotypic viabilities on co-existing genotypes in *Drosophila*. *Heredity, Lond.* **13**, 393–402.

—— (1959*b*). Genotypic-environmental interactions for various temperatures in *Drosophila melanogaster*. *Genetics, Princeton* **44**, 1325–33.

—— (1961). Fly size, emergence time and sternopleural chaeta number in *Drosophila*. *Heredity, Lond.* **16**, 455–73.

—— (1963*a*). A widespread biochemical polymorphism in *Drosophila melanogaster*. *Am. Nat.* **97**, 375–82.

—— (1963*b*). Migration as a factor in natural selection. *Genetica* **33**, 184–206.

—— (1963*c*). Polymorphism and the balanced polygenic complex. *Evolution, Lancaster, Pa.* **17**, 564–74.

—— (1963*d*). Complex polymorphisms where the coupling and repulsion double heterozygote viabilities differ. *Heredity, Lond.* **18**, 369–74.

—— (1964). A diallel cross for mating speeds in *Drosophila melanogaster*. *Genetica* **35**, 141–51.

—— (1965*a*). The determination of mating speeds in *Drosophila melanogaster* for various combinations of inbred lines. *Experientia* **21**, 478.

—— (1965*b*). Assortative mating for a metrical characteristic in *Drosophila*. *Heredity, Lond.* **20**, 161–7.

—— (1966). The genotypic control of longevity in *Drosophila melanogaster* under two environmental regimes. *Aust. J. biol. Sci.* **19**, 587–91.

—— (1967*a*). *The genetic analysis of behaviour*. Methuen, London.

—— (1967*b*). Behaviour and random mating. *Experientia* **23**, 585.

—— (1968). Genetic heterogeneity among the founders of laboratory populations of *Drosophila melanogaster*. III. Sternopleural chaetae. *Aust. J. biol. Sci.* **21**, 297–302.

—— (1969). A correlation between the ability to withstand high temperatures and radioresistance in *Drosophila melanogaster*. *Experienta* **25**, 1000.

—— (1970*a*). Genetic heterogeneity in natural populations of *Drosophila melanogaster* for ability to withstand dessication. *Theoret. appl. Genet.* **40**, 261–6.

—— (1970*b*). Genetic heterogeneity among the founders of laboratory populations of *Drosophila melanogaster*. V. Sternopleural and abdominal chaetae in the same strains. *Theoret. Appl. Genet.* **40**, 337–40.

—— (1971). Extreme-environment heterosis and genetic loads. *Heredity, Lond.* **26**, 579–83.

—— and GREEN, M. M. (1959). Pleiotropy and competition at the vermilion locus in *Drosophila melanogaster*. *Proc. natnl. Acad. Sci. U.S.A.* **45**, 993–6.

—— and HOSGOOD, S. M. W. (1967). Genetic heterogeneity among the founders of laboratory populations of *Drosophila*. I. Scutellar chaetae. *Genetica* **38**, 328–39.

—— ——, and LEE, B. T. O. (1967). Polygenes and polymorphism. *Molec. Gen. Genetics* **99**, 165–70.

—— and KAUL, D. (1966). Mating speed and duration of copulation in *Drosophila pseudoobscura*. *Heredity, Lond.* **21**, 219–25.

—— and KROMAN, R. A. (1960). Melanin inhibitors and the ebony locus in *Drosophila melanogaster*. *Heredity, Lond.* **15**, 301–14.

——, MACBEAN, I. T., and LEE, B. T. O. (1969). Polymorphism in natural populations for genes controlling radioresistance in *Drosophila*. *Genetics, Princeton* **61**, 211–18.

—— and MCKENZIE, J. A. (1972). The ecological genetics of *Drosophila*. *Evolutionary Biol.*

PATTERSON, J. T. and STONE, W. S. (1952). Evolution in the genus *Drosophila*. Macmillan, New York.

PEREZ-SALAS, S., RICHMOND, R. C., PAVLOVSKY, O., KASTRITSIS, C. D., EHRMAN, L., and DOBZHANSKY, T. (1970). The Interior semispecies of *Drosophila paulistorum*. *Evolution, Lancaster, Pa.* **24**, 519–27.

PERTTUNEN, V. and AHONEN, U. (1956). The effect of age on the humidity reaction of *Drosophila melanogaster* (Dipt. Drosophilidae). *Suom. hyönt. Aikak.* **22**, 63–71.

—— and SALMI, H. (1956). The responses of *Drosophila melanogaster* (Dipt. Drosophilidae) to the relative humidity of the air. *Suom. hyönt. Aikak.* **22**, 36–45.

PETIT, C. (1951). Le rôle de l'isolement sexuel dans l'évolution des populations de *Drosophila melanogaster*. *Bull. Biol. Fr. Belg.* **85**, 392–418.

—— (1954). L'isolement sexuel chez *Drosophila melanogaster*. Étude du mutant *white* et son allélomorph *sauvage*. *Bull. Biol. Fr. Belg.* **88**, 435–43.

—— (1958). Le déterminisme génétique et psycho-physiologique de la compétition sexuelle chez *Drosophila melanogaster*. *Bull. Biol. Fr. Belg.* **92**, 248–329.

—— and EHRMAN, L. (1969). Sexual selection in *Drosophila*. In *Evolutionary biology* vol. 3. (eds. T. Dobzhansky, M. K. Hecht, and W. C. Steere), pp. 177–223. Appleton-Century-Crofts, New York.

PHAFF, H. J., MILLER, M. W., RECCA, J. A., SHIFRINE, M., and MRAK, E. M. (1956). Studies on the ecology of *Drosophila* in the Yosemite region of California. II. Yeasts found in the alimentary canal of *Drosophila*. *Ecology* **37**, 533–8.

PHILIP, U., RENDEL, J. M., SPURWAY, H., and HALDANE, J. B. S. (1944). Genetics and karyology of *Drosophila subobscura*. *Nature* **154**, *Lond.* 260–2.

PIELOU, E. C. (1966). Shannon's formula as a measure of specific diversity: its use and misuse. *Am. Nat.* **100**, 463–5.

—— (1969). *An introduction to mathematical ecology*. Wiley-Interscience, New York.

PIPKIN, S. B. (1953). Fluctuations in *Drosophila* populations in a tropical area. *Am. Nat.* **87**, 317–22.

—— (1965). The influence of adult and larval food habits on population size of neotropical ground-feeding *Drosophila*. *Am. Midl. Nat.* **74**, 1–27.

——, RODRÍGUEZ, R. L., *and* LEÓN, J., (1966). Plant host specificity among flower-feeding neotropical *Drosophila* (Diptera: Drosophilidae). *Am. Nat.* **100**, 135–56.

PITTENDRIGH, C. S. (1958). Adaptation, natural selection, and behavior. In *Behavior and evolution* (eds. A. Roe and G. G. Simpson), pp. 390–416. Yale University Press, New Haven.

PRAKASH, S. (1967). Association between mating speed and fertility in *Drosophila robusta*. *Genetics, Princeton* **57**, 655–63.

—— (1968). Chromosome interactions affecting mating speed in *Drosophila robusta*. *Genetics, Princeton* **60**, 589–600.

—— *and* LEWONTIN, R. C. (1968). A molecular approach to the study of genic heterozygosity in natural populations. III. Direct evidence of coadaptation in gene arrangements of *Drosophila*. *Proc. natnl. Acad. Sci. U.S.A.* **59**, 398–405.

—— ——, *and* HUBBY, J. L. (1969). A molecular approach to the study of genic heterozygosity in natural populations. IV. Patterns of genic variation in central, marginal and isolated populations of *Drosophila pseudoobscura*. *Genetics, Princeton*, **61**, 841–58.

PREVOSTI, A. (1955). Geographical variability in quantitative traits in populations of *Drosophila subobscura*. *Cold Spring Harb. Symp. quant. Biol.* **20**, 294–9.

—— (1966). Chromosomal polymorphism in western Mediterranean populations of *Drosophila subobscura*. *Genet. Res.* **7**, 149–58.

PROUT, T. (1971*a*). The relation between fitness components and population prediction in *Drosophila*. I. The estimate of fitness components. *Genetics, Princeton* **68**, 127–49.

—— (1971*b*). The relation between fitness components and population prediction in *Drosophila*. II. Population prediction. *Genetics, Princeton* **68**, 151–67.

RASMUSON, B. (1955). A nucleo-cytoplasmic anomaly in *Drosophila melanogaster* causing increased sensitivity to anaesthetics. *Hereditas* **41**, 147–208.

RENDEL, J. M. (1945). Genetics and cytology of *Drosophila subobscura*. II. Normal and selective matings in *Drosophila subobscura*. *J. Genet.* **46**, 287–302.

—— (1951). Mating of ebony vestigial and wild type *Drosophila melanogaster* in light and dark. *Evolution, Lancaster, Pa.* **5**, 226–30.

RICHARDSON, R. H. (1969). Migration, and enzyme polymorphisms in natural populations of *Drosophila*. *Jap. J. Genet.* **44**, suppl. 1, 172–9.

RICHMOND, R. C. *and* POWELL, J. R. (1970). Evidence of heterosis associated with an enzyme locus in a natural population of *Drosophila*. *Proc. natnl. Acad. Sci. U.S.A.* **67**, 1264–7.

RISCH, P. (1971). Die überwinterung als adaptive leistung, geprüft an verschiedenen karyotypen der inversionspolymorphen art *Drosophila subobscura*. *Genetica* **42**, 79–103.

ROBERTS, R. C. (1967). Some concepts and methods in quantitative genetics. In *Behavior-genetic analysis* (ed. J. Hirsch), pp. 214–57. New York, McGraw-Hill.

ROBERTSON, F. W. (1960*a*). The ecological genetics of growth in *Drosophila*. I. Body size and developmental time on different diets. *Genet. Res.*, **1**, 288–304.

—— (1960*b*). The ecological genetics of growth in *Drosophila*. *Genet. Res.* **1**, 305–18.

—— (1962). Changing the relative size of the body parts of *Drosophila* by selection. *Genet. Res.* **3**, 169–80.

ROBERTSON, F. W. (1963). The ecological genetics of growth in *Drosophila*. 6. The genetic correlation between the duration of the larval period and body size in relation to larval diet. *Genet. Res.* **4**, 74–92.

—— (1966). The ecological genetics of growth in *Drosophila*. 8. Adaptation to a new diet. *Genet. Res.* **8**, 165–79.

——, SHOOK, M., TAKEI, G., *and* GAINES, H. (1968). Observations on the biology and nutrition of *Drosophila disticha*. Hardy, an indigenous Hawaiian species. *Univ. Tex. Publs.* No. 6818, 279–99.

SAKAI, K.-I. (1961). Competitive ability in plants: its inheritance and some related problems. *Symp. Soc. exp. Biol.* **15**, 245–63.

——, NARISE, T., HIRAIZUMI, Y., *and* IYAMA, S.-Y. (1958). Studies on competition in plants and animals. IX. Experimental studies on migration in *Drosophila melanogaster*. *Evolution Lancaster, Pa.* **12**, 93–101.

SALT, R. W. (1961). Principles of insect cold-hardiness. *A. Rev. Ent.* **6**, 55–74.

SAMEOTO, D. D. *and* MILLER, R. S. (1966). Factors controlling the productivity of *Drosophila melanogaster* and *D. simulans*. *Ecology* **47**, 695–704.

SAMMETA, K. P. V. *and* LEVINS, R. (1970). Genetics and ecology. *A. Rev. Genet.* **4**, 469–88.

SANG, J. H. (1956). The quantitative nutritional requirements of *Drosophila melanogaster*. *J. exp. Biol.* **33**, 45–72.

SCHULTZ, J. *and* REDFIELD, H. (1951). Interchromosomal effects on crossing over in *Drosophila*. *Cold Spring Harb. Symp. quant. Biol.* **16**, 175–97.

SCOTT, J. P. (1943). Effects of single genes on the behavior of *Drosophila*. *Am. Nat.* **77**, 184–90.

SEXTON, O. J. *and* STALKER, H. D. (1961). Spacing patterns of female *Drosophila paramelanica*. *Anim. Behav.* **9**, 77–81.

SHOREY, H. H. (1962). Nature of the sound produced by *Drosophila melanogaster* during courtship. *Science, N.Y.* **137**, 677–8.

—— *and* BARTELL, R. J. (1970). Role of a volatile female sex pheromone in stimulating male courtship behaviour in *Drosophila melanogaster*. *Anim. Behav.* **18**, 159–64.

SIEGEL, I. M. (1967). Heritability and threshold determinations of the optomotor response in *Drosophila melanogaster*. *Anim. Behav.* **15**, 299–306.

SOKAL, R. R., EHRLICH, P. R., HUNTER, P. E., *and* SCHLAGER, G. (1960). Some factors affecting pupation site of *Drosophila*. *Ann. ent. Soc. Am.* **53**, 174–82.

SOKOLOFF, A. (1955). Competition between sibling species of the *pseudoobscura* subgroup of *Drosophila*. *Ecol. Monogr.* **25**, 387–409.

—— (1964). Studies on the ecology of *Drosophila* in the Yosemite region of California. V. A preliminary survey of species associated with *D. pseudoobscura* and *D. persimilis* at slime fluxes and banana traps. *Pan-Pacif. Ent.* **40**, 203–18.

—— (1966). Morphological variation in natural and experimental populations of *Drosophila pseudoobscura* and *Drosophila persimilis*. *Evolution, Lancaster, Pa.* **20**, 49–71.

SOLAR, E. DEL (1968). Selection for and against gregariousness in the choice of oviposition sites by *Drosophila pseudoobscura*. *Genetics, Princeton* **58**, 275–82.

—— *and* PALOMINO, H. (1966). Choice of oviposition sites in *Drosophila melanogaster*. *Am. Nat.* **100**, 127–33.

SOUTHWOOD, T. R. E. (1966). *Ecological methods*. Methuen, London.

SOUZA, H. M. L. DE, CUNHA, A. B. DA, *and* SANTOS, E. P. DOS (1970). Adaptive polymorphism of behavior evolved in laboratory populations of *Drosophila willistoni*. *Am. Nat.* **104,** 175–89.

SPASSKY, B. (1951). Effect of temperature and moisture content of the nutrient medium on the viability of chromosomal types in *Drosophila pseudoobscura*. *Am. Nat.* **85,** 177–80.

—— and DOBZHANSKY, T. (1967). Responses of various strains of *Drosophila pseudoobscura* and *Drosophila persimilis* to light and to gravity. *Am. Nat.* **101,** 59–63.

—— SPASSKY, N., LEVENE, H., *and* DOBZHANSKY, T. (1958). Release of genetic variability through recombination. I. *Drosophila pseudoobscura*. *Genetics, Princeton* **43,** 844–67.

SPIESS, E. B. (1958). Effects of recombination on viability in *Drosophila*. *Cold Spring Harb. Symp. quant. Biol.* **23,** 239–50.

—— (1959). Release of genetic variability through recombination. II. *Drosophila persimilis*. *Genetics, Princeton* **44,** 43–58.

—— (1968a). Courtship and mating time in *Drosophila pseudoobscura*. *Anim. Behav.* **16,** 470–9.

—— (1968b). Low frequency advantage in mating of *Drosophila pseudoobscura* karyotypes. *Am. Nat.* **102,** 363–79.

—— (1970). Mating propensity and its genetic basis in *Drosophila*. In *Essays in evolution and genetics in honor of Theodosius Dobshansky* (eds. M. K. Hecht and W. C. Steere), pp. 315–79. Appleton-Century-Crofts, New York.

—— and LANGER, B. (1961). Chromosomal adaptive polymorphism in *Drosophila persimilis*. III. Mating propensity of homokaryotypes. *Evolution, Lancaster, Pa.* **15,** 535–44.

—— —— (1964a). Mating speed control by gene arrangements in *Drosophila pseudoobscura* homokaryotypes. *Proc. natnl. Acad. Sci. U.S.A.* **51,** 1015–19.

—— —— (1964b). Mating speed control by gene arrangement carriers in *Drosophila persimilis*. *Evolution, Lancaster, Pa.* **18,** 430–44.

—— ——, *and* SPIESS, L. D. (1966). Mating control by gene arrangements in *Drosophila pseudoobscura*. *Genetics, Princeton* **54,** 1139–49.

—— and SPIESS, L. D. (1967). Mating propensity, chromosomal polymorphism, and dependent conditions in *Drosophila persimilis*. *Evolution, Lancaster, Pa.* **21.** 672–8.

—— —— (1969). Mating propensity, chromosomal polymorphism, and dependent conditions in *Drosophila persimilis*. II. Factors between larvae and between adults. *Evolution, Lancaster, Pa.* **23,** 225–36.

SPIESS, L. D. *and* SPIESS, E. B. (1969). Minority advantage in interpopulational matings of *Drosophila persimilis*. *Am. Nat.* **103,** 155–72.

SPIETH, H. T. (1951). The breeding site of *Drosophila lacicola* Patterson. *Science, N.Y.* **113,** 232.

—— (1952). Mating behavior within the genus *Drosophila* (Diptera). *Bull. Am. Mus. nat. Hist.* **99,** 401–74.

—— (1958). Behavior and isolating mechanisms. In *Behavior and evolution* (eds. A. Roe and G. G. Simpson), pp. 363–89. Yale University Press, New Haven.

—— (1968). Evolutionary implications of sexual behavior in *Drosophila*. *Evolutionary Biol.* **2,** 157–93.

—— and HSU, T. C. (1950). The influence of light on the mating behavior of seven species of the *Drosophila melanogaster* species group. *Evolution, Lancaster, Pa.* **4,** 316–25.

STALKER, H. D. (1942). Sexual isolation studies in the species complex *Drosophila virilis*. *Genetics, Princeton* **27**, 238–57.
—— and CARSON, H. L. (1948). An altitudinal transect of *Drosophila robusta* Sturtevant. *Evolution, Lancaster, Pa.* **2**, 295–305.
—— —— (1949). Seasonal variation in the morphology of *Drosophila robusta* Sturtevant. *Evolution, Lancaster, Pa.* **3**, 330–43.
STONE, W. S., GUEST, W. C., *and* WILSON, F. D. (1960). The evolutionary implications of the cytological polymorphism and phylogeny of the *virilis* group of *Drosophila*. *Proc. natnl. Acad. Sci. U.S.A.* **46**, 350–61.
——, KOJIMA, K.-I., *and* JOHNSON, F. M. (1969). Enzyme polymorphisms in animal populations. *Jap. J. Genet.* **44**, suppl. 1, 166–71.
STRICKBERGER, M. W. *and* WILLS, C. J. (1966). Monthly frequency changes of *Drosophila pseudoobscura* third chromosome gene arrangements in a California locality. *Evolution, Lancaster, Pa.* **20**, 592–602.
STURTEVANT, A. H. (1915). Experiments on sex recognition and the problem of sexual selection in *Drosophila*. *J. anim. Behav.* **5**, 351–66.
—— (1929). The genetics of *Drosophila simulans*. *Publs. Carnegie Instn.* No. 399, 1–62.
—— *and* BEADLE, G. W. (1936). The relations of inversions in the X chromosome of *Drosophila melanogaster* to crossing over and disjunction. *Genetics, Princeton* **21**, 554–604.
SVED, J. A. (1968). The stability of linked systems of loci with a small population size. *Genetics, Princeton* **59**, 543–63.
——, REED, T. E., *and* BODMER, W. F. (1967). The number of balanced polymorphisms that can be maintained in a natural population. *Genetics, Princeton* **55**, 469–81.
SZEBENYI, A. L. (1969). Cleaning behaviour in *Drosophila melanogaster*. *Anim. Behav.* **17**, 641–51.
TANTAWY, A. O. (1964). Studies on natural populations of *Drosophila*. III. Morphological and genetic differences of wing length in *Drosophila melanogaster* and *D. simulans* in relation to season. *Evolution, Lancaster, Pa.* **18**, 560–70.
—— *and* EL-WAKIL, H. M. (1970). Studies on natural populations of *Drosophila*. XI. Fitness components and competition between *Drosophila funebris* and *D. virilis*. *Evolution, Lancaster, Pa.* **24**, 528–30.
—— *and* MALLAH, G. S. (1961). Studies on natural populations of *Drosophila*. I. Heat resistance and geographical variation in *Drosophila melanogaster* and *D. simulans*. *Evolution, Lancaster, Pa.* **15**, 1–14.
——, MOURAD, A. M., *and* MASRI, A. M. (1970). Studies on natural populations of *Drosophila*. VIII. A note on the directional changes over a long period of time in the structure of *Drosophila* near Alexandria, Egypt. *Am. Nat.* **104**, 105–9.
THIESSEN, D. D., OWEN, K., *and* WHITSETT, M. (1970). Chromosome mapping of behavioral activities. In *Contributions to behavior-genetic analysis: The mouse as a prototype* (eds. G. Lindzey and D. D. Thiessen), pp. 161–204. Appleton, New York.
THODAY, J. M. (1961). Location of polygenes. *Nature* **191**, 368–70.
—— (1964). Genetics and the integration of reproductive systems. In *Insect reproduction* (ed. K. C. Highnam), *Symp. R. Ent. Soc.*, No. 2, pp. 108–19.
—— *and* BOAM, T. B. (1961). Regular responses to selection. I. Description of responses. *Genet. Res.* **2**, 161–76.

THOMSON, J. A. (1971). Association of karyotype with body weight and resistance to desiccation in *Drosophila pseudoobscura*. *Can. J. Genet. Cytol.* **13**, 63–9.

TIMOFEEFF-RESSOVSKY, N. W. (1940). Mutations and geographical variation. In *The new systematics* (ed. J. Huxley), pp. 73–136. Oxford University Press.

—— and TIMOFEEFF-RESSOVSKY, E. A. (1941a). Populationsgenetische Versüche an *Drosophila*. I. Zeitliche und räumliche Verteilung der Individuen einiger *Drosophila*—Arten über das Gelande. *Z. indukt. Abstamm.- u. Vererb. Lehre* **79**, 28–34.

—— —— (1941b). Populationsgenetische Versuche an *Drosophila*. II. Aktionsbereiche von *Drosophila funebris* und *Drosophila melanogaster*. *Z. indukt. Abstamm.- u. Vererb.Lehre* **79**, 35–43.

TOBARI, I. (1966). Effects of temperature on the viabilities of homozygotes and heterozygotes for second chromosomes of *Drosophila melanogaster*. *Genetics, Princeton* **54**, 783–91.

TOWNSEND, J. I. (1952). Genetics of marginal populations of *Drosophila willistoni*. *Evolution, Lancaster, Pa.* **6**, 428–42.

TSUKAMOTO, M. (1955). Mode of inheritance of resistance to nicotine sulfate in *Drosophila melanogaster*. *Botyu-Kagaku* **20**, 73–81.

—— and OGAKI, M. (1954). Gene analysis of resistance to DDT and BHC. *Botyu-Kagaku* **19**, 25–32.

VAN VALEN, L., LEVINE, L., and BEARDMORE, J. A. (1962). Temperature sensitivity of chromosomal polymorphism in *Drosophila pseudoobscura*. *Genetica* **33**, 113–27.

VETUKHIV, M. (1953). Viability of hybrids between local populations of *Drosophila pseudoobscura*. *Proc. natnl. Acad. Sci. U.S.A.* **39**, 30–4.

—— (1954). Integration of the genotype in local populations of three species of *Drosophila*. *Evolution, Lancaster, Pa.* **8**, 241–51.

—— (1956). Fecundity of hybrids between geographic populations of *Drosophila pseudoobscura*. *Evolution, Lancaster, Pa.* **10**, 139–46.

—— (1957). Longevity of hybrids between geographic populations of *Drosophila pseudoobscura*. *Evolution, Lancaster, Pa.* **11**, 348–60.

VOLTERRA, V. (1926). Variazioni e fluttuazioni del numero d'individui in specie animali conviventi. *Atti Accad. naz. Lincei Memorie*, ser. 6., **2**, 31–113. (English translation: Variations and fluctuations of the number of individuals in animal species living together. In R. N. Chapman, 1931, *Animal ecology*, pp. 409–48. McGraw-Hill, New York.)

WADDINGTON, C. H. (1959). Canalization of development and genetic assimilation of acquired characters. *Nature, Lond.* **183**, 1654–5.

——, WOOLF, B., and PERRY, M. M. (1954). Environment selection by *Drosophila* mutants. *Evolution, Lancaster, Pa.* **8**, 89–96.

WALLACE, B. (1954). Genetic divergence of isolated populations of *Drosophila melanogaster*. *Carylogia* **6**, suppl. vol., 761–4.

—— (1966a). On the dispersal of *Drosophila*. *Am. Nat.* **100**, 551–63.

—— (1966b). Distance and the allelism of lethals in a tropical population of *Drosophila melanogaster*. *Am. Nat.* **100**, 565–78.

—— (1966c). The fate of sepia in small populations of *Drosophila melanogaster*. *Genetica* **37**, 29–36.

—— (1968). *Topics in population genetics*. Norton, New York.

WALLACE, B. *and* VETUKHIV, M. (1955). Adaptive organization of the gene pools of *Drosophila* populations. *Cold Spring Harb. Symp. quant. Biol.* **20**, 303–10.

WEARDEN, S. (1964). Alternative analyses of the diallel cross. *Heredity, Lond.* **19**, 669–80.

WEISBROT, D. R. (1966). Genotypic interactions among competing strains and species of *Drosophila*. *Genetics, Princeton* **53**, 427–35.

WHITE, M. J. D. (1956). Adaptive chromosomal polymorphism in an Australian grasshopper. *Evolution, Lancaster, Pa.* **10**, 298–313.

—— (1958). Restrictions on recombination in grasshopper populations and species. *Cold Spring Harb. Symp. quant. Biol.* **23**, 307–17.

—— (1962). Genetic adaptation. *Aust. J. Sci.* **25**, 179–86.

WIGAN, L. G. (1944). Balance and potence in natural populations. *J. Genet.* **46**, 150–60.

WILLIAMS, C. M. *and* REED, S. C. (1944). Physiological effects of genes: The flight of *Drosophila* considered in relation to gene mutations. *Am. Nat.* **78**, 214–23.

WILLIAMSON, D. L. *and* EHRMAN, L. (1967). Induction of hybrid sterility in nonhybrid males of *Drosophila paulistorum*. *Genetics, Princeton* **55**, 131–40.

WOLKEN, J. J., MELLON, A. D., *and* CONTIS, G. (1957). Photoreceptor structures. II. *Drosophila melanogaster*. *J. exp. Zool.* **134**, 383–406.

WRIGHT, S. (1921). Systems of mating. *Genetics, Princeton* **6**, 111–78.

—— (1968). Dispersion of *Drosophila pseudoobscura*. *Am. Nat.* **102**, 81–4.

—— *and* DOBZHANSKY, T. (1946). Genetics of natural populations. XII. Experimental reproduction of some of the changes caused by natural selection in certain populations of *Drosophila pseudoobscura*. *Genetics, Princeton* **31**, 125–56.

—— ——, *and* HOVANITZ, W. (1942). Genetics of natural populations. VII. The allelism of lethals in the third chromosome of *Drosophila pseudoobscura*. *Genetics, Princeton* **27**, 363–94.

AUTHOR INDEX

Bold type indicates a page on which a literature reference is given

Ahonen, U., 93, **210**
Akihama, T., 111, **193**
Anderson, W. W., 80, 81, 85, 99, 142, 150, **193, 198**
Andrewartha, H. G., 129, 134, **193**
Ayala, F. J., 128, 131–5, 138, 169, 173–7, 191, **193**

Bakker, K., 112, **194**
Ball, F. M., 140, **207**
Band, H. T., 146–7, **194**
Barker, J. S. F., 26, 173, 180, **194**
Barnes, B. W., 99, **194, 204**
Bartell, R. J., 79, **212**
Bastock, M., 11–15, 18, 19, 27–9, 57, 77, 183, **194**
Bateman, A. J., 23, 120, **194**
Battaglia, B., 138, **194**
Beadle, G. W., 136, **214**
Beardmore, J. A., 88, 131, 137–9, 146, 148, 152, 160, **194, 197, 215**
Beck, J., 28, **196**
Becker, H. J., 76, **194**
Bennet-Clark, H. C., 180, 182, **194**
Benzer, S., 29, **194**
Berger, E. M., 160–1, **194**
Birch, L. C., 110, 129–30, 134, 140, **193, 195**
Blight, W. C., 167, **195**
Boam, T. B., 62, **214**
Bodmer, W. F., 144, 150, 153, 156, **195, 214**
Borowsky, R., 177, **195**
Bösiger, E., 51, **195**
Boudreau, J. C., 75, **203**
Breese, E. L., 106, **195**
Brenton, W., 98, **201**
Brncic, D., 37, 92, 96, 137, 141, 164–5, **195**
Broadhurst, P. L., 43, 49, **195**
Brown, R. G. B., 85, 183, **195**
Budnik, M., 137, **195**
Burla, H., 119, 144, **195, 199**
Burnet, B., 27–8, **196, 197**
Burns, J. M., 160–1, **196**

Carmody, G., 83–4, **196**
Carpenter, F. W., 76, **196**
Carson, H. L., 3–5, 92, 96–8, 122–3, 131, 142, 145, 163–7, 178–9, 183–5, 189–90, **196, 199, 214**

Caspari, E., 8, 60, **196**
Castro, L. E., 146, **203**
Cavalcanti, A. G. L., 119, **195**
Cavalli-Sforza, L. L., 54, 121, **196**
Chejne, A. J., 86, **203**
Chetverikov, S. S., 2, **196**
Church, R. B., 108, **197**
Clausen, J., 2, **197**
Clayton, F. E., 4, **197**
Collazo, A. D., 83–4, **196**
Connolly, K., 27, 28, 59, 75, 77, **196, 197**
Contis, G., 76, **216**
Cooch, F. G., 88, **197**
Cooke, P., 111, **207**
Cooper, D. M., 162–3, 178, **197, 199**
Cordeiro, A. R., 147, **197**
Crossley, S., 183, **197**
Crow, J. F., 100–2, 128, 150, 152, 189, **197**
Crumpacker, D. W., 95–6, 153, **197, 207**
Cunha, A. B. da, 119, 131, 136, 139, 144, 163, **195, 197, 198, 199, 212**

Darlington, C, D., 145, **198**
Dawood, M. M., 113, 138, **198**
Deery, B. J., 103–4, **198**
Demerec, M., 1, **198**
Dobzhansky, T., 2, 5, 31, 66, 75–6, 79, 82–5, 118–22, 126–31, 134, 136–45, 147–8, 150, 152–3, 162–3, 174, 178–9, 182, **193, 194, 195, 196, 197, 198, 199, 200, 201, 204, 206, 207, 210, 213, 216**
Dubinin, N. P., 141–2, **200**

Ehrlich, P. R., 172, **212**
Ehrman, L., 23, 35, 79–81, 83–5, 87, 88, 126, 182, 190, **193, 196, 199, 200, 201, 210, 216**
Elens, A. A., 15, 21, 26, **201**
Elliott, P. O., 129–30, 134, **195**
El-Wakil, H. M., 173, **214**
Epling, C., 140, **207**
Erlenmeyer-Kimling, L., 63, 66, **201, 203**
Erway, L., 98, **201**
Ewing, A. W., 12–13, 72, 178, 180, 182, **194, 201**

Falconer, D. S., 43, 58, 62, **201**
Fine, W., 157, **208**
Fingerman, M., 29, **201**

AUTHOR INDEX

Fisher, R. A., 31, 43, 89, 124, 134, 156, 188–91, **201**
Fitzpatrick, T. B., 103, **205**
Fontdevila, A., 153, **201**
Ford, E. B., 2, **201**
Franklin, I., 157–9, **201**
Fraser, A., 98, **201**
Freire-Maia, N., 168, **201**
Fulker, D. W., 53, 54, 60, 188, **201**
Fuller, J. L., 2, **201**
Futuyma, D. J., 173, **201**

Gaines, H., 165, **212**
Gale, J. S., 111–12, 115, **202**
Gause, G. F., 175–7, **202**
Geer, B. W., 27, **202**
Georghiou, G. P., 100–1, **202**
Gibson, J. B., 88, 154, 160, **202**
Gillaspy, J. F., 160, **204**
Gillespie, J., 156–7, **205**
Glass, B., 124, **202**
Granobles, L. A., 146, **203**
Grant, B., 75, **202**
Green, M. M., 27–8, **202, 210**
Griffing, B., 49, 51, 105, 153, **202, 206**
Grigolo, A., 102, **202**
Grossfield, J., 26, **202**
Grove, J., 167, **205**
Guest, W. C., 5, **214**

Hadler, N. M., 75–6, **202**
Haldane, J. B. S., 26, 144, 152, **202, 210**
Halpern, E. A., 96, **206**
Hardy, D. E., 3–5, 92, 163–5, 183–5, **196**
Harlan, H. V., 110, **202**
Harrison, B. J., 63, 87, **207**
Hay, D. A., 54–7, 60, 77, **202**
Hayman, B. I., 49, 54, **202**
Heed, W. B., 96, 163, 165–7, **202, 204, 205, 206**
Herskowitz, I. H., 103, **202**
Heuts, M. J., 92, 96, 152, **203**
Hiesey, W. M., 2, **197**
Hiraizumi, Y., 124, **212**
Hirsch, J., 63–6, 75, **201, 203**
Hoenigsberg, H. F., 86, 146, 167, **203**
Hortobagji-German, E., 86, **203**
Hosgood, S. M. W., 67, 90–1, 106, 189, **203, 210**
Hovanitz, W., 121–2, **216**
Hsu, T. C., 26, **213**
Huang, S. L., 116, **203**
Hubby, J. L., 147, 153, 161, **206, 211**
Hunter, A. S., 147, **199**
Hunter, P. E., 172, **212**
Hutchinson, G. E., 168, **203**

Idrobo, J. M., 146, **203**

Ives, P. T., 96, 146–7, **194, 203**
Iyama, S.-Y., 124, **212**

Jacobs, M. E., 27, **203, 204**
Jaeger, E. C., 147, **197**
Jaffrey, I. S., 83–4, **196**
Jinks, J. L., 43, 48–9, 54, **204, 207**
Johnson, F. M., 155, 159–61, **196, 204, 214**

Kalmus, H., 30, 92, **204**
Kambysellis, M. P., 165–6, **204**
Kaplan, W. D., 29, **204**
Kastritsis, C. D., 83, 182, **199, 210**
Kato, S. K., 140, **207**
Kaul, D., 32–5, 38, 74, 152, **204, 210**
Kearsey, M. J., 68, 69, 99, 105–6, 112, **204**
Keck, D. D., 2, **197**
Kempthorne, O., 43, 49, **204**
Kessler, S., 48, 54, 73, 180–1, **204**
Kimball, S., 83–4, **196**
Kimura, M., 152, 189, **197**
King, J. C., 102, **204**
Kircher, H. W., 165, 167, **202, 205**
Knapp, E. P., 163, 178, **196, 199**
Knight, G. R., 86, **205**
Koehn, R. K., 160, **205**
Kojima, K-I., 68–9, 105–6, 116, 155–7, 159, **203, 204, 205, 214**
Koopman, K. F., 179–80, **205**
Koref-Santibañez, S., 37, 86, 137, 182, **195, 205**
Kosuda, K., 150, **205**
Krimbas, C. B., 141, 148, **205**
Kroman, R. A., 103, **205, 210**

Lack, D., 134, **205**
Langer, B., 26, 30–2, 34–7, 56, 188, **213**
Langridge, J., 153–5, 186, **205**
Lee, B. T. O., 9, 63, 66, 90, 106, 157–8, **205, 210**
León, J., 165, **211**
Lerner, A. B., 103, **205**
Levene, H., 23, 143–4, 148, **199, 205, 206, 213**
Levine, L., 137, 152, **215**
Levine, R. P., 93, **206**
Levins, R., 95, 98–9, 110, 115, 169–71, **206, 212**
Lewontin, R. C., 26, 76, 112–14, 116, 129–130, 134, 139, 147, 153, 157–9, 161, **195, 199, 201, 205, 206, 211**
Li, C. C., 124, **202**
Long, T., 139, **206**
Lowe, C. H., 96, **206**

MacArthur, R. H., 145, 168, 170–1, **206**
MacBean, I. T., 44–6, 66–70, 90, **206, 210**
McFarquhar, A. M., 150, **206**

MacIntyre, R. J., 157, 161, **208**
McKenzie, J. A., 15–19, 21, 24, 38, 168, **206, 210**
McWilliam, J. R., 153, **206**
Mainardi, M., 171, **206**
Mallah, G. S., 90–1, 98–9, 169, **214**
Malogolowkin-Cohen, C. H., 23, **206**
Manning, A., 28, 37, 70–5, 180, **206**
Marinkovic, D., 95–6, 153, **197, 207**
Martini, M. L., 110, **202**
Masri, A. M., 168, **214**
Mather, K., 43, 48–9, 54, 63, 87–8, 106, 111–12, 117, 157–9, 187, **195, 207**
Matsuo, Y., 112–14, 116, 139, **206**
Mayhew, S. H., 140, **207**
Maynard Smith, J., 51, 88, **207**
Mayr, E., 57, 82, 88, 179, 189, **199, 207**
Médioni, J., 29, 76, **207, 208**
Mellon, A. D., 76, **216**
Merrell, D. J., 24–5, 28, 34, 102, 172, 188, **208**
Mettler, L. E., 75, **202**
Miklovich, R., 154, **202**
Miller, D. D., 82, **208**
Miller, M. W., 163, **210**
Miller, R. S., 172, **212**
Misra, R. K., 97, **208**
Miyoshi, Y., 108, **208**
Mohn, N., 96, **208**
Moore, J. A., 172–4, **208**
Mourad, A. M., 168, **214**
Mrak, E. M., 163, **210**
Muller, H. J., 152, **208**
Murakami, S., 104–5, **208**

Nair, P. S., 147, **208**
Nakashima-Tanaka, E., 90, 104–5, **208**
Narise, T., 124–6, **208, 212**

Obrebski, S., 83–4, **196**
O'Brien, S. J., 157, 161, **208**
O'Donald, P., 88, **208**
Ogaki, M., 90, 100, 104–5, **208, 215**
Ogita, Z., 101–2, **208, 209**
Ohba, S., 130, **209**
Oka, H. I., 111, **209**
Oppenoorth, F. J., 100, 102, **202, 209**
Oshima, C., 102, 142, 146, **193, 209**
Owen, K., 28, **214**

Palomino, H., 171, **212**
Parsons, P. A., 7, 9, 15–19, 21–3, 28, 32–5, 38, 43–6, 48–53, 56, 60, 63, 66, 67, 74, 88–91, 93, 95, 98–9, 103–4, 106–7, 114–15, 123–4, 144, 150, 152–4, 156–9, 168, 189, **195, 198, 203, 204, 205, 206, 209, 210**
Patterson, J. T., 5, **210**
Pavan, C., 119, 162, **199**

Pavlovsky, O., 79, 83, 85, 129–31, 134, 138, 142–4, 147, 153, **193, 194, 198, 199, 201, 206, 210**
Perez-Salas, S., 83–4, **210**
Perry, M. M., 92, **215**
Perttunen, V., 92–3, **210**
Petersen, J. A., 147, **197**
Petit, C., 23, 78–9, 126, **210**
Phaff, H. J., 163, 178, **210**
Philip, U., 26, **210**
Pielou, E. C., 170–1, **210**
Pipkin, S. B., 162, 165, **210**
Pittendrigh, C. S., 26, 76, 92, **211**
Podger, R. N., 173, **194**
Powell, J. R., 155, **211**
Prakash, S., 37, 54, 147, 161, **211**
Prevosti, A., 97, 145, **211**
Prout, T., 187–8, **211**

Rasmuson, B., 104, **211**
Rasmussen, D. I., 160, **205**
Recca, J. A., 163, **210**
Redfield, H., 145, **212**
Reed, S. C., 13, **216**
Reed, T. E., 153, **214**
Reeve, E. C. R., 97, **208**
Rendel, J. M., 26–7, **210, 211**
Richardson, R. H., 120, 122, 126, **211**
Richmond, R. C., 155, **210, 211**
Risch, P., 96, **211**
Roberts, R. C., 43, **211**
Robertson, A., 86, **205**
Robertson, F. W., 97, 103, 108–9, 114, 50, 165, **197, 206, 211, 212**
Rockwood, E. S., 160, **204**
Rodríguez, R. L., 165, **211**
Romano, A., 167, **195**
Russell, J. S., 167, **205**

Sakai, K.-I., 110–11, 124, **212**
Salceda, V. M., 96, 153, **207**
Salmi, H., 92, **210**
Salt, R. W., 96, **212**
Sameoto, D. D., 172, **212**
Sammeta, K. P. V., 110, 115, **212**
Sang, J. H., 108, **212**
Santos, E. P. dos, 131, 139, **212**
Schaffer, H. F., 160, **204**
Schlager, G., 172, **212**
Schultz, J., 145, **212**
Scott, J. P., 29, 76, **212**
Sewell, D., 27–8, **197**
Sexton, O. J., 55, **212**
Shifrine, M., 163, **210**
Shook, M., 165, **212**
Shorey, H. H., 79, 179, **212**
Siegel, I. M., 76, **212**
Silagi, S., 83–4, **196**

Simmons, A. S., 23, **206**
Singh, M., 116, **203**
Smith, H., 138, **194**
Sokal, R. R., 172, **212**
Sokoloff, A., 98, 110, 163, 167, **198, 212**
Solar, A. del, 171, **212**
Sømme, J. C., 102, **204**
Southwood, T. R. E., 21, **212**
Souza, H. M. L., de, 131, 139, **212**
Spassky, B., 66, 75, 76, 79, 83, 93, 142, 144, 147–8, 150, 153, **198, 199, 201, 213**
Spassky, N., 148, 153, **199, 213**
Spiess, E. B., 20, 25–6, 31–2, 34–8, 56, 79, 92, 96, 148, 188, **208, 213**
Spiess, L. D., 34–8, 79, 92, **213**
Spieth, H. T., 3–5, 26, 85, 92, 163–5, 167, 179, 183–5, **196, 213**
Spurway, H., 26, **210**
Stalker, H. D., 22, 55, 97–8, 142, **212, 214**
Stone, W. S., 3–5, 92, 155, 159, 163–5, 183–185, **196, 210, 214**
Streisinger, G., 82, **200**
Strickberger, M. W., 113, 138, 142, **198, 214**
Sturtevant, A. H., 11, 24, 136, 180, **214**
Sved, J. A., 153, 156, **214**
Szebenyi, A. L., 77, **214**

Takei, G., 165, **212**
Tantawy, A. O., 90–1, 98–9, 168–9, 173, **214**
Thiessen, D. D., 28, **214**
Thoday, J. M., 9, 62–3, 66, 88, **202, 214**
Thompson, W. R., 2, **201**
Thomson, J. A., 93–4, 152, **214**
Tidwell, T., 83–4, **196**
Timofeeff-Ressovsky, E. A., 119, **215**
Timofeeff-Ressovksy, N. W., 92, 118–19, **215**
Tiniakov, G. G., 141–2, **200**
Tobari, I., 154, **215**

Tobari, Y. N., 156–7, **205**
Townsend, J. I., 147, **197, 215**
Trout, W. E., 29, **204**
Tsukamoto, M., 100, **215**

Ullrich, R., 83–4, **196**
Underhill, J. C., 102, **208**

van Valen, L., 137, 152, **215**
Vetukhiv, M., 87, 149, **215, 216**
Volterra, V., 176, **215**

Waddington, C. H., 86, 92, 108, **205, 215**
Wallace, B., 86, 91–2, 118–23, 126, 137, 147, 149, 152, 155, 168, 172, **199, 215, 216**
Wasserman, M., 166, **196**
Watanabe, T., 142, **193, 209**
Wattiaux, J. M., 21, **201**
Wearden, S., 49, **216**
Weisbrot, D. R., 113, **216**
Weiss, J. M., 63, 66, **201**
Westphal, N. J., 82, **208**
Wheeler, M. R., 167, **196**
White, M. J. D., 136, 145, **216**
Whitsett, M., 28, **214**
Wigan, L. G., 150, **216**
Williams, C. M., 13, **216**
Williamson, D. L., 84, **216**
Wills, C. J., 142, **199, 214**
Wilson, F. D., 5, **214**
Wolken, J. J., 76, **216**
Woolf, B., 92, **215**
Wright, S., 31, 43, 118–22, 126, 128, 136, 138, 152, **200, 216**

Yarbrough, K. M., 155, **205**

Zuill, E., 183, **197**

SUBJECT INDEX

acclimation, 95, 189
activity, 54–5, 59, 71–2, 75, 178
adaptedness, 127–9, 133–5, 174, 191
adaptive radiation, 5, 163, 189
alcohol dehydrogenase, 154, 156, 160
allopatric species, 83–5, 182
anaesthetics, ether, 104–5
  chloroform, 105
Arabidopsis thaliana, 153
Arctic skua, 88
assortative mating, 59, 78, 88–9

balanced genotype, 157–9, 187
barley, 110–11
biometrical genetics, 9, 41–60
Blue Snow Goose, 88
body weight (size), 89, 93, 95, 98–9, 103, 108, 114, 149

Catostomus clarkii, 160
central populations, 131, 133, 144–8, 166, 189
Cereus giganteus, 96
Cestrum parqui, 164
Cheirodendron, 163–5
chemical labelling of flies, 120, 122, 126
chemical stresses, 100–9
  genetic architectures, 106–7
chemotaxis, 76–7
choice experiments, 15–22, 24–9, 80–9
choice indices, 22–4
Chusquea, 165
Cibotium, 164
cleaning behaviour, 77
Clermontia, 163–4
clines, altitudinal, 140–1
  geographical, 97–9, 124, 141–2, 159–60
coadaptation, 142–4, 148–51, 158, 187
community diversity, 171
competition, within species, 110–17, 140
  between species, 172–7
conditioned media, 113–14, 116
co-operation, 112–13, 117
copulation duration, 33–4, 44–6, 66–70
courtship, 11–15, 25, 72–3, 79, 83, 178, 180, 182–5
courtship songs, 178–80, 182
Cucumis melo, 120

Datura arbustiva, 165
DDT, 100–2, 107, 142, 190–1

density-dependence, *see* frequency-dependence
desiccation, 92–5, 162, 189
diallel cross, 9, 48–57, 104–5
directional selection, 9–10, 54–5, 61–77, 100–3, 106–7, 111, 190–1
dispersion, 118–26, 189
disruptive (diversifying) selection, 61, 75, 88
diurnal behaviour, 178, 183, 185
Drosophila
  aldrichi, 120
  allei, 165
  americana, 167
  ananassae, 159, 162
  anuda, 162
  appendiculata, 165
  athabasca, 82
  auraria, 26–7
  azteca, 163
  bifasciata, 182
  birchii, 129, 132–5
  bucksii, 112–13, 116, 123, 139, 163, 165
  buzzattii, 124, 166
  californica, 167
  carcinophila, 167
  cardini, 169–70
  disticha, 165
  endobranchia, 167
  equinoxalis, 79
  flavopilosa, 92, 96, 141, 164
  funebris, 92, 119–20, 141–2, 172–3
  gaucha, 37, 86, 182
  hydei, 123, 163, 169–70
  hypocausta, 162
  imaii, 182
  immigrans, 123, 163
  melanogaster, 1, 2, 7–20, 24–30, 34, 44–6, 49–57, 59, 62–73, 75–9, 86–96, 98–105, 107–9, 111–16, 119, 121–6, 135, 138–9, 145–50, 153–7, 160–3, 167–74, 180, 182–183, 187–8
  miranda, 182
  montana, 167
  nebulosa, 172–4
  nigrospiracula, 96
  pachea, 166–7
  palustris, 26
  paulistorum, 82–5, 143, 155, 182
  pavani, 36, 86, 137, 141, 182
  pavlovskiana, 83

SUBJECT INDEX

Drosophila—contd.
  persimilis, 26, 31, 35–8, 56, 66, 76, 79, 86, 92, 96, 148, 163, 167, 178–82
  prosaltans, 82, 95, 148
  pseudoobscura, 2, 7, 8, 26, 31–8, 48, 54, 56, 66, 73–6, 79–81, 85–7, 92–6, 98–9, 113, 118–20, 124, 128–31, 134–5, 136–145, 147–50, 152–3, 161, 163, 167, 171–182, 186–8
  repleta, 169–70
  robusta, 37, 54, 96–7, 131, 133, 141, 144–5, 147, 167
  serrata, 129, 132–5, 138, 172–7
  simulans, 15, 26, 73–4, 86, 90–1, 98, 123, 161, 167–8, 172–4, 180, 182–3
  sturtevanti, 82
  subobscura, 26–7, 51, 96–7, 141, 145, 150, 182
  tropicalis, 79
  victoria, 167
  virilis, 101, 173
  willistoni, 79, 82, 95, 119–20, 131, 136, 139, 141, 143–5, 147–9, 169–70, 172
duration of copulation, 33–4, 44–6, 66–70
dysprosium acetate, 120

ecological isolation, 178–80, 183
environmental effects, behaviour, 26–7, 36–38, 55–7, 76
  equilibria, 136–42, 152–5, 186
  morphology, 97–100
  population size, 132–5
environmental stresses, 90–109
  chemical, 100–9
  desiccation, 92–5, 162, 169
  temperature, 90–6, 152–5, 186
enzyme variants, 2–3, 147, 153–7, 159–61, 186
escape behaviour, 75
Escherichia coli, 154
ethanol, 168, 180
extreme environments, *see* environmental stresses
eye pigments, effect on mating, 27–8

fecundity potential, 165–6
fitness, 127–9, 143, 149, 186–92
  definition, 127–8
  in fluctuating environments, 139
  relations between components, 32, 114–115, 187–9
food sources, 162–8
founder effect, 90, 123, 189–90
frequency-dependence, 40
  competition, 112–17, 177
  mating, 57, 78–80, 188
  polymorphisms, 155

fundamental theorem of natural selection, 134, 188–9

Gause's principle, 175–7
Gecarcinus lateralis, 167
  ruricola, 167
genetic architecture, 105–7, 157–9, 189
genetic background, 15, 29, 123, 125, 150
genetic load, 128–9, 147, 152–9
genotype–environment interaction, 44, 46, 48, 55–7, 172
geotaxis, 63–6, 75

Hawaiian Drosophilidae, 3–5, 163–6, 183–5, 189, 192
Hemiargus isola, 160
heritability, 46–7, 58, 62
heterogamic matings, 22–3, 83, 86
heterosis, 31–7, 39–40, 136–44, 149–50, 152–5, 186–8
homogamic matings, 20, 22–3, 83–6
humidity preferences, 92–3, 164, 178, 184–5
hybrid sterility, 82, 84, 179, 182

innate capacity for increase, 129–30, 174, 187
insecticides, 100–2, 107, 142, 190–1
inversions, altitudinal variations, 141–2
  behaviour, 31–8, 56, 66
  marginal and central populations, 144–148
  seasonal variations, 139–40
  temperature, 136–42, 152, 186 (*see also* co-adaptation)
isolation, *see* ecological *and* sexual
isolation indices, 22–4, 83–4, 87–8

karyotypes,
  D. persimilis, 35–6
  D. pseudoobscura, 31–2, 145
  D. robusta, 144–5
  D. willistoni, 144–5
  Hawaiian Drosophilidae, 4–5, 189
  marginal and central populations, 144–6

lethal genes, 121–2, 146–8, 150
light, effect on mating, 12, 26–7
  intensity, 164, 184–5
linkage disequilibria, 156–9
living space, 134–5
Lophocereus schottii, 166

maize, 153
marginal populations, 131, 133, 144–8, 166, 189
mating chamber, 16, 18, 21–2, 26, 79
mating frequency (propensity, speed), 11–22, 24–9, 31–8, 45, 48–54, 70–4, 187–8

migration, 118–26, 189–90
  laboratory studies, 124–6
monophagous species, 165
morphological traits, 97–100
mutual facilitation, 113–14, 138

NaCl, 107–8, 139
neurological mutants, 29
niche breadth, 168–72
nutritional requirements, 108–9, 165–8

objectivity in measurement, 6–7
olfactory factors, 15, 57, 79
optometer response, 30, 76
Opuntia, 124, 166
overdominance, *see* heterosis
over-wintering, 96, 147
oviposition site, 163–5, 171–2, 180

Penicillium, 112
pheromone, 79
phototaxis, 26, 29, 75–6
pilocereine, 167
Pisonia, 164
Pogonomyrex barbatus, 160
polygene, 42
polymorphism, 2, 31–40, 80, 90, 128, 130–1, 136–48, 152–61, 186–7
polyphagous species, 165
population size, experimental populations, 127–35, 187
  natural populations, 138–9, 146–8, 162
preening, 54–7, 77
principle of competitive exclusion, 175–7
productivity, 131–5, 187
PTU (phenylthiourea), 100, 102–3, 114
pupation site, 131, 139, 172

quantitative genetics, 9, 41–60
Quercus kelloggii, 167, 179

random mating, 16, 22, 39, 78, 87
regression of offspring on parent, 58–9
rice, 111

Salix interior, 166
Scaptomyza, 3–5
scaptomyzoids, 3–5, 164

schottenol, 166–7
seasonal changes, inversion frequencies, 92, 140–2
  morphological, 97–8
  species distributions, 162–3
selection, 9, 61–77
  behavioural analysis, 70–7
  DDT resistance, 101–2, 190–1
  genetic analysis, 63–70
  responses, 62
  sexual isolation, 86–9, 179–81
sexual dimorphism, 184–5
sexual isolation, 22, 78, 80–9, 178–82
sexual selection, 22–5, 27–8, 78, 80–1
sibling species, 86, 163, 167–8, 172–3, 178–182
slime flux, 167, 179
species distributions, 162–85
stabilizing selection, 61, 75, 100, 106, 157–9
supercooling, 96
sympatric species, 84–5, 178, 182–3

temperature, fluctuations, 91–2, 139, 146, 168–9, 180
  high, 90–2, 95–6, 114, 152–5, 189
  low, 92, 95–6, 153–5
  preferences, 163–4, 178, 180, 184
temperature effects, equilibria, 136–8, 140, 160, 186
  interspecific competition, 173–6
  mating frequencies, 36–8
  population size, 130, 132–5
territoriality, 184–5
Thrips tabaci, 164

Ulmus americana, 167

variance, 40–1
  additive genetic, 47
  dominance, 47
  environmental, 44–7
  epistatic, 47
  genotypic, 44–7
  phenotypic, 44–7
visual stimuli, 12, 26

wheat, 111

yeasts, 141, 163, 165